AF597893

CHILDREN'S ENCYCLOPEDIA

OUR PLANET EARTH

ANIMALS | HUMAN BODY

(An imprint of Prakash Books)

contact@wonderhousebooks.com

Disclaimer: The information contained in this encyclopedia has been collated with inputs from subject experts. All information contained herein is true to the best of the Publisher's knowledge.

ISBN : 978-93-58567-00-7

Table of Contents

OUR PLANET EARTH

Planet Earth's rich and diverse features make it a truly exceptional place, providing a habitat for countless organisms and a remarkable environment for exploration and discovery. It is home to stunning mountains, lush forests, expansive grasslands, hot deserts, and vast oceans that are teeming with life.

Mountains tower over the earth's surface and can reach dizzying heights, with majestic peaks and stunning, snow-capped summits. Forests abound across the planet, nurturing an incredible range of flora and fauna, while also providing critical benefits such as oxygen production, carbon sequestration, and water regulation.

Grasslands stretch for vast expanses across the planet and support a rich variety of wildlife, including large herds of grazing herbivores such as bison, giraffes, and antelopes. Deserts, on the other hand, often form in the world's driest regions with harsh conditions that have unique adaptations to surviving the heat and scarcity of water.

Rocks and minerals, and the soils derived from them, are the foundation of our planet. They make up the Earth's crust, and the mineral and nutrient composition of soil influences the health, growth, and diversity of plant life.

The oceans that cover over 70% of the planet's surface are massive and diverse ecosystems, home to a vast range of animal species, from tiny plankton to giant whales. They also play a crucial role in regulating the Earth's climate and providing key resources such as food and energy.

Volcanoes and earthquakes are some of the most dynamic and powerful geologic events on our planet. Volcanoes spew molten lava and ash, shaping the landscape over thousands of years, while earthquakes happen when plates in the Earth's crust shift and move, causing ground-shaking and sometimes tsunamis.

Weather and climate are vital components of the Earth's system, with weather playing a crucial role in day-to-day life and climate influencing ecosystems and human societies on a larger scale. The complex interactions of the oceans, atmosphere, and landform the foundation of weather and climate, leading to diverse and fascinating consequences.

Together, all these natural wonders and phenomena make the planet earth a marvel of complexity and beauty.

▼ *The elegant snow-capped mountains and the green forest cover are two examples of Earth's diverse ecosystems*

Majestic Mountains

Mountains proudly rise above the Earth's crust, both on land and water. They have steep sloping sides, rounded ridges, and a summit. Mountains are steeper, larger, and taller than hills and rise at least 1,000 metres above the surrounding landforms. They rise next to each other forming a chain that is called a mountain range.

Making Mountains

The Earth's crust is made up of large slabs of rocks called tectonic plates. These plates fit into each other like the pieces of a jigsaw puzzle.

Mountains are formed along the boundaries of these tectonic plates when they move towards each other.

While moving, they collide with each other, causing a kind of deformity in the crust. This leads to a crustal uplift and formation of a mountain. This process happens horizontally, causing the crust to fold into several layers and form wrinkles along the convergent plate boundaries. These wrinkles eventually become mountains. It takes millions of years for a mountain to be formed.

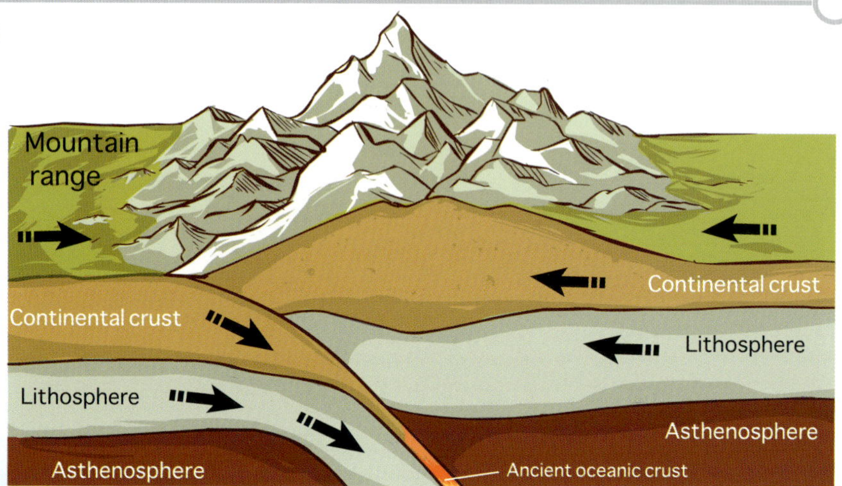

The diagram shows the tectonic plates converging and forming a mountain

Convergent Plate Boundaries

Tectonic plates form convergent plate boundaries when they move towards each other. They can form a continental-oceanic convergence, oceanic-oceanic convergence, or continental-continental convergence.

- **Continental-oceanic convergence:** The oceanic plate moves underneath the continental plate. This process is called **subduction**. Volcanoes and earthquakes are formed through this process.
- **Continental-continental convergence:** When two continental plates converge, this collision leads to the formation of big mountains. This type of convergence formed the Himalayan mountain range.
- **Oceanic-oceanic convergence:** When two oceanic plates converge, the older, denser plate sinks below the other. This creates a subduction zone, forming a chain of curved volcanic mountains called island arcs.

The Aleutian mountain range in Alaska wa formed by oceanic-oceanic convergence

Fault Lines

Mountains are also formed along natural fault lines. These mountains form when faults or cracks in Earth's crust force some materials or blocks of rock upwards. In this case, instead of the crust folding over, it fractures and pulls apart. Due to this immense force, blocks of rock along the sides of these faults are lifted up and tilted sideways.

The Vosges Mountains in France and the Black Forest mountain range in Germany are fault mountains separated by the Rhine Valley.

Types of Mountains

Mountains differ in their appearance and formation. They can be classified based on the types of rocks, shape and placement on land. From the perspective of formation, they can be divided into these types: fold mountains, fault-block mountains, volcanic mountains, dome mountains, and plateau mountains.

Fold Mountains

Fold mountains are formed when two tectonic plates collide with each other. Slowly, the fold that forms from this movement takes the shape of a mountain. Many of the world's great mountain ranges like the Himalayas in Asia, the Andes in South America, the Rocky Mountains in North America, the Alps in Europe, and the Ural Mountains in Russia fall under this category. Fold mountains are the youngest mountains formed on Earth.

▲ *Mount Kangchenjunga, the third highest peak in the Himalayan mountain range at 8,586 metres, is a fold mountain*

◀ *Another name for volcanic mountains is mountains of accumulation. Mauna Kea and Mount Fuji are volcanic mountains*

Volcanic Mountains

Volcanic mountains are formed when magma comes out onto Earth's surface. At the surface, the magma erupts as lava, ash, rocks, and volcanic gases. They fall around the vent in successive layers, like a volcanic cone, that pile up and form a mountain. The best examples of volcanic mountains are Mauna Loa and Mount Kea on the Big Island of Hawaii; Mount Fuji in Japan, Mount Mayon in Philippines, and Mount Merapi in Sumatra are great examples of volcanic mountains.

Dome Mountains

Dome mountains are formed when a large amount of magma pushes its way upwards from beneath Earth's crust. Even before it can erupt, the source of magma goes away, leaving the pushed-up rocks as they were. These rocks then cool and form a mountain. With time, it takes the shape of a dome. The Black Hills and the Adirondack Mountains in the USA are examples of dome mountains.

▼ *Dome mountains are not found in mountain belts; instead they are isolated and individual structures*

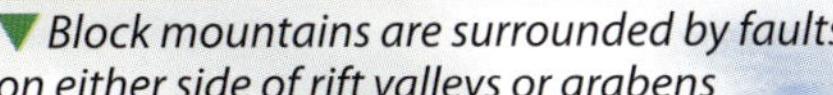

▼ *Block mountains are surrounded by faults on either side of rift valleys or grabens*

Fault-block Mountains

Fault-block mountains are formed along faults where some large blocks of rock are forced upwards, while others are pushed downwards. As a result, instead of folding, they break up into chunks. The Sierra Nevada Mountains in the USA and the Harz Mountains in Germany are examples of fault-block mountains.

Plateau Mountains

Plateau mountains are generally formed by erosion and not because of the activities taking place beneath Earth's crust. These mountains are found near fold mountains. North Island Volcanic Plateau in New Zealand and the Catskill Mountains of the USA are examples of plateau mountains.

▲ *Plateau mountains can rise more than 600 metres above the sea level*

Other Landforms On Earth

Landforms occur naturally on Earth's crust. The regal mountains are one such landform. Other landforms like hills, plateaus, plains, glaciers, valleys, and deserts add to the natural beauty of our planet and make it a marvellous place to live in. All these landforms have their own special characteristics and have a great impact on the environment and life on Earth.

Landforms are created due to the movement of tectonic plates inside Earth's crust. Some are formed by erosion—the movement of water and wind. They wear down the land and create landforms like valleys and canyons. Both processes happen over a long period of time, sometimes taking millions of years.

Hills

A hill is a part of land that is elevated above everything that surrounds it. It is shorter than a mountain and less steep. However, like a mountain, a hill has an obvious summit, which is its highest point. Like mountains, there are different types of hills. Butte is another type of hill. These are hills that usually stand alone in a flat area.

▲ *Hills are also formed by erosion and deposition by wind and water*

In Real Life

Earlier, hills that were 300 metres or higher were considered to be mountains. However, geologists who study landforms have now put forth a ruling that hills only above 1,000 metres will be considered as mountains.

Plateaus

A plateau is an elevated, flat-topped landform that rises sharply above the surrounding area. The word 'plateau' is derived from the French word 'platel', which means flat or plate. It is usually bounded by steep slopes on all sides and is sometimes enclosed by mountains. They are different from mountains as they have a totally flat top. These landforms are found in all the continents.

There are two types of plateaus—Dissected plateaus and Volcanic plateaus.

▼ *Plateaus are formed in the same way as mountains and hills, that is, by the movement of tectonic plates or by volcanic activity*

Isn't It Amazing!

The Colorado Plateau is a dissected plateau that lies in the western part of the USA. It has been rising by .03 centimetres every year, for more than 10 million years.

Valleys

A valley is a depression in the Earth's surface, between hills or mountains. Valleys are formed due to erosion caused by rivers. This process takes millions of years. They all take the form of a 'U' or a 'V'. The kind of valley formed is defined by the type of geological process it has gone through. There are three main types of valleys—river valleys, glacier valleys, and rift valleys.

Isn't It Amazing!

There is another kind of 'valley' called a dell, which is a small, wooded or partially-wooded hollow. Dells that are covered with ravines or trees are called dingles. In Scotland, a dell is called a glen.

▲ *The walls of the 'Iao Valley are covered with forests. The gorge is 8 kilometres long*

▲ *Trekking in the Barun Valley is tough along the valley, but the scenic beauty of the Himalayas makes it worth it.*

▲ *The Moraine Lake is unique due to its blue colour and stunning reflections of snow-capped mountain*

▲ *People riding the double-humped camel found in this part of India. They are natives of Mongolia in central Asia and were brought to Nubra Valley during the golden era of trade on the Silk Route*

Plains

Plains are broad areas of flat land. They cover one-third of the world's area and are present on all continents on Earth. Entire civilisations were built on plains because early hunters and gatherers found fertile land for agricultural purposes, giving them a chance to settle down. Plains are also formed by the movement of fast-flowing rivers.

There are many types of plains. These are the flood plains, alluvial plains, coastal plains, and abyssal plains. Plains do not show much elevation. They occur alongside of rivers, coastlines, bottoms of valleys, and next to plateaus.

▼ *Plains are suitable for agriculture*

▲ *Much of the Great Plains are covered in prairies, steppes, and grasslands*

Our Ecosystem

What is an ecosystem? It includes all the living things (biotic)—such as plants, animals, and other organisms—interacting with each other, and with the non-living things (abiotic)—such as the weather, the Sun, soil, water, and air—in a particular region.

Biomes

The whole surface of Earth is a series of connected ecosystems which are again connected in a larger biome. A biome is a large region of Earth that has a particular type of climate and living organisms. The major biomes of Earth include forests, grasslands, and deserts, and each one has many ecosystems. The animals and plants within have particular characteristics that help them survive in that biome.

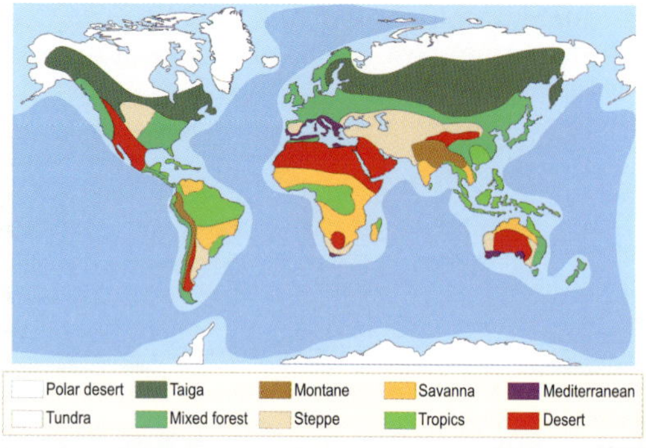

The map shows the major biomes in the world, such as the desert, tundra, taiga, etc.

Forests: Not Just a Collection of Trees

A forest is a complex ecosystem that consists of many different types of trees, plants, and animals. The trees help sustain life on Earth as they give out oxygen that human beings and animals need to survive. They are an important component of the environment as they have the ability to clean the air, cool it on hot days, conserve heat at night, and act as excellent sound absorbers. Forests can be divided into three types—tropical forests, temperate forests, and **boreal** forests, which are also known as 'taiga.'

Tropical Rainforests

Tropical rainforests are found in areas that receive a considerable amount of rainfall, mostly in wet tropical uplands and lowlands around the Equator.

Location

Tropical rainforests are mostly found in South and Central America, West and Central Africa, Eastern Madagascar and the Zaire basin, Indonesia, parts of Southeast Asia, and tropical Australia. Some of the largest rainforests in the world are the Amazon Rainforest spread over South America, the Congo Rainforest in Central Africa, and the Sundaland in South East Asia.

Climate

The climate in the tropical rainforest regions is humid with very little seasonal variation.

Flora and Fauna

The tropical rainforests contain more than two-thirds of the world's flora species. Different animals live in different parts of the rainforest. For example, large animals live on the forest floor, birds live in tree canopies or at the tops of taller trees, arboreal animals live on trees, and insects are found everywhere.

Isn't It Amazing!

80 per cent of the flowers that are found in the Australian rainforests are unique and cannot be found anywhere else in the world.

Temperate Forests

Temperate forests are found between the tropics and the polar regions, within the temperate zone. Temperate forests are primarily deciduous, with tall, broad-leafed hardwood trees that shed their leaves and change colour during autumn. There are also **coniferous** trees in these regions such as cypress, cedar, redwood, fir, juniper, and pine trees. These trees have needle-like leaves and cones.

Location

Temperate forests are dominant in large areas of North America and Eurasia and in smaller portions of the Southern Hemisphere. Temperate deciduous forests are found in eastern USA, Canada, Europe, China, Japan, and western Russia.

Climate

Temperate forests exist in regions having moderate climate. They are not too hot like the tropical rainforests, or too cold like the boreal forests. They experiences the four seasons—summer, autumn, winter, and spring.

▲ *Leaves change colour because the chlorophyll is consumed by the trees as the climate changes*

Fauna

Wildlife in these forests either endures the winter or migrates to warmer climates. As the days become shorter and temperatures drop during fall.

Animals such as slugs, frogs, turtles, and salamanders are common in temperate forests. Birds such as broad-winged hawks, snowy owls, and pileated woodpeckers are also found here. Apart from these animals, mammals such as the white-tailed deer, raccoons, opossums, porcupines, and red foxes are quite common in temperate forests.

▲ *The mona monkey is found in lowland forests of eastern Ghana, Togo, Benin, Nigeria, and western Cameroon*

Flora

Temperate forests have a great variety of plant species. Small plants like lichen, moss, ferns, and wildflowers are found on the forest floor. Shrubs fill in the middle level. Temperate forests are dominated by deciduous trees such as oak, maple, and beech. Conifers like cedar, spruce, fir, and pine trees can also be found mixed in with the hardwood trees in this biome.

▶ *Deodar, a kind of cedar, is mostly found in the western Himalayas of India*

Evergreen Forests

Temperate forests can be broadly divided into evergreen and deciduous forests. Evergreen forests are made up of evergreen trees. The word 'evergreen' means trees that do not shed their leaves completely. Different trees shed their leaves at different times of the year, thus such forests always appear to be green.

Location

They are found in all the continents, except Antarctica. Tropical evergreen forests are found in the regions where there is ample rainfall and the temperature is 15–30° C. Evergreen trees are found all over the world, especially in the cold regions of the Northern Hemisphere.

▶ *Evergreen trees can grow almost anywhere, but some types cannot thrive in cooler climates*

Soil and Sunlight

Evergreen trees usually require a lot of sunlight, but there are exceptions. Though they need an excess of rainfall to grow, some evergreen trees can thrive in conditions of draught. Coniferous evergreen trees require dry and well-drained soil, whereas rainforests can adapt to moist and soggy soil.

Deciduous Forests

Deciduous is a word derived from the Latin word 'decidere', which means 'tending to fall off.' Therefore, deciduous forests have trees and plants that shed their leaves most often in autumn and regain them in spring.

▲ *Autumn in a deciduous forest*

Location

Deciduous forests are found almost all over the planet in both the hemispheres. However, the world's largest deciduous forests are seen in the Northern Hemisphere, in North America, Europe, parts of Russia, China, and Japan. In the Southern Hemisphere, deciduous forests are found in some parts of Australia, southern Asia, and South America.

Temperature and Precipitation

The temperature of each region varies, depending on its location. The average temperature of a deciduous forest is usually around 10° C, with winters being much colder. Deciduous forests require ample rain to promote new leaf growth on trees.

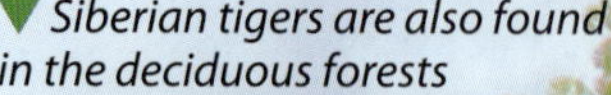

▼ *Siberian tigers are also found in the deciduous forests*

Flora and Fauna of Deciduous Forests

Deciduous forests are home to many different species of plants. Some common plants of deciduous forests are oak, maple, spruce, pine, birch, and beech etc. Some unique inhabitants are the orchids and rhododendrons. Shrubs such as lichen and moss are quite common in these forests.

Taiga

Taiga forests are located to the south of the tundra and north of the temperate forests. It is in the subarctic region, lying to the south of the Arctic Circle.

Location

The regions in the world that have taiga forests are Alaska, Canada, Scandinavia, and Siberia. Russia has the largest stretch of taiga forests from the Pacific Ocean to the Ural Mountains. Taiga forests occupy 27 per cent of Earth's surface.

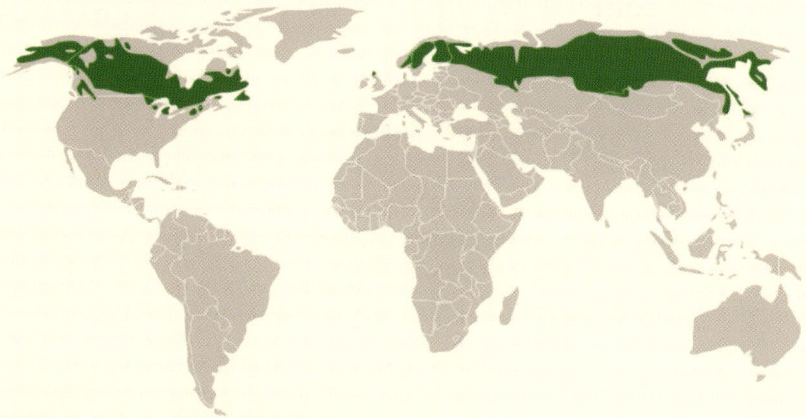

▲ *The map marks the taiga regions*

Climate

The taiga biome is very close to the Arctic Circle and is extremely cold with temperatures mostly below 0° C. Throughout the year, the area experiences freezing cold weather.

Flora

Due to the cold climate, there is less diversity in the type of vegetation here. The trees here are conifers like spruce, pine, fir, cedar, hemlock, and larches and the boreal forest is a wooded biome.

The taiga is covered with mosses, lichens, and mushrooms. The plants in this region have developed special key features to adapt to the varying climate of the taiga biome. They can grow directly on the ground and have very shallow roots.

Fauna

The animals, like the plants, need to adapt to these tough conditions to survive. Animals found here are mammals such as bears, deer, moose, elk, caribou, ermines, moles, squirrels, chipmunks, and bobcats. Herbivorous animals are seen in areas that have a greater number of trees. Many species of birds such as sparrows, finches, woodpeckers, crows, and eagles are also seen here, along with a variety of species of insects.

Antarctica

The continent of Antarctica is permanently covered with snow. A mere 1 per cent is left for plants to grow.

Climate

The climate of Antarctica is extremely cold and dry. The temperature in winter, along the coast, ranges from -10 to -30° C. However, in the mountainous regions, the temperature drops below -60° C.

Fauna

It is home to numerous animal species. The most familiar animals are penguins, whales, seals, albatrosses, and a variety of marine life in the ocean.

Flora

Only two species of flowering plants are found, the Antarctic hair grass and Antarctica pearlwort in the South Orkney Islands. Vegetation here consists of mosses, liverworts, lichens, and fungi which has adapted to the conditions, especially the dryness.

Grassy Grasslands

Grasslands are areas where different species of grasses dominate the vegetation. They are the most agriculturally useful habitats on Earth. Grasslands are found in regions where the rainfall received is not enough to support the growth of forests.

Types of Grasslands

Tropical Grasslands

Tropical grasslands are found mainly near the Equator, between the Tropic of Cancer and the Tropic of Capricorn. They are present in Africa, as well as in large areas of Australia, South America, and India. These grasslands are warm and have a dry and a rainy season.

Characteristics

The tropical grasslands have more woody shrubs than the temperate ones. During the rainy season, flowers bloom, and some survive till the winter as they have thick stems to store water.

Animals of the Grasslands

There are many animals native to the tropical grasslands. They include giraffes, zebras, buffaloes, kangaroos, mice, moles, gophers, ground squirrels, snakes, worms, termites, beetles, lions, leopards, hyenas, and elephants.

Temperate Grasslands

Temperate grasslands are mainly found in the belt between the forests and deserts on the Great Plains of North America. Similar to the savannas, temperate grasslands are areas of open grasslands with very few trees. However, they are located in colder regions and receive little precipitation. These are located to the north of the Tropic of Cancer and south of the Tropic of Capricorn.

Rainfall in Temperate Grasslands

Temperate grasslands receive low to moderate precipitation. This is the reason tall trees and large shrubs are not found here. Droughts and fires restrict trees from growing. There are some 100 species of flowers that grow among the grasses.

Wildlife in Temperate Grasslands

Wildlife is abundant in temperate grasslands. They are home to many large herbivores such as bisons, gazelles, zebras, rhinoceroses, and wild horses. There are also carnivores such as lions and wolves in the temperate grasslands. Some smaller animals are also found here, such as deer, prairie dogs, mice, jackrabbits, skunks, coyotes, snakes, foxes, badgers, and grasshoppers.

In Real Life

▲ *Sometimes, the burning of sawdust and dry leaves due to friction results in spontaneous fires in the grasslands*

Periodic fires occur in the grasslands and are quite common. This is required to destroy unwanted plants that tend to encroach into the grassland. Fires can be induced by human beings or they may start spontaneously. Forest fires do not burn down the grassland plants as their buds are below the ground. Animals are also protected by fire as many live underground. Other animals and birds are mobile enough to cope with the problem.

▼ *Grasslands are also called pampas in South America and steppes in Eurasia. They have tough plants that can tolerate freezing temperatures, droughts, fires, and grazing animals*

▶ *Grasses that dominate this ecosystem with a lot of widely spaced trees are termed as savannas*

Rolling in the Deserts

A desert is dry land with sparse vegetation. Deserts cover around 20 per cent of Earth's surface. They are found in all the continents. A desert is a place that receives less than 25 centimetres of rainfall per year. It is an ecosystem that supports plants and animals with special characteristics required to survive in a tough environment.

▲ *The different deserts of the world, located within each continent*

Location

Deserts are generally found around the Tropics of Cancer and Capricorn. Deserts are mainly located towards the west of different continents. This is mainly due to the easterly winds of the tropics, which are also known as the trade winds. There are four types of deserts based upon their geographic situation—polar deserts, subtropical deserts, cold winter deserts, and cool coastal deserts.

The world's largest desert is the Sahara Desert in the Northern Hemisphere. Other large deserts are the Arabian Peninsula in West Asia, and the Thar Desert in India and Pakistan. There is a cold desert in Central Asia known as the Gobi Desert. Some deserts of the Southern Hemisphere are Atacama and Patagonia in South America, Namib and Kalahari in Africa, and the Great Sandy Desert in Australia.

Climate

The climate of deserts is usually hot and dry as there is very little precipitation. In the hot deserts, the day temperatures are often above 38° C in summer. However, there is a drastic fall in the temperature at night. The fluctuations in the temperature and scarcity of water make the deserts a harsh place to live.

Flora

In deserts, plants do not get water for a long period of time. Therefore, there are certain plants with special roots that help them absorb the small amount of water that is available.

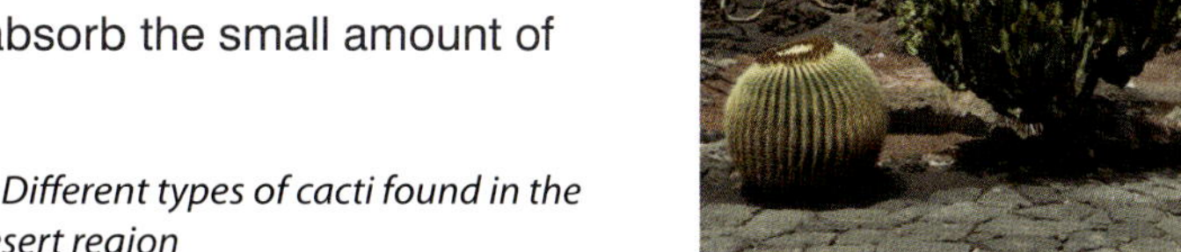

▶ *Different types of cacti found in the desert region*

Fauna

Camels can go for days without food and water, and their nostrils and eyelashes protect them from the sand. There are animals that come out in the night when the temperature falls. Animals like the desert tortoise in south-western USA spend much of their time underground. Birds in the desert region are nomadic.

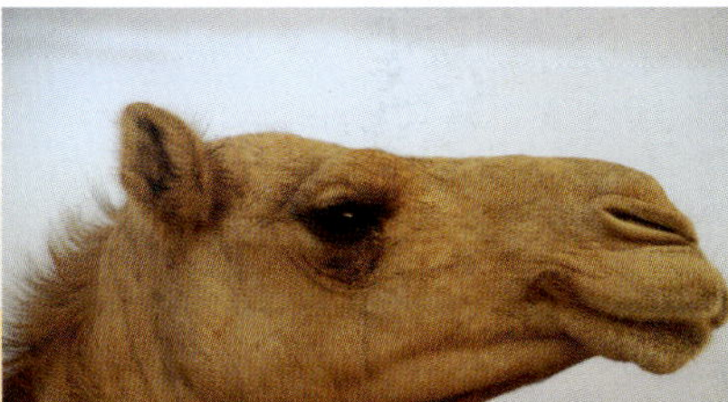

◀ *Camels are smarter than horses, and they pee on their back legs to help cool them down*

What is an Ocean?

Oceans make a continuous body of salt water covering 70 per cent of Earth's surface. Oceans contain almost 98 per cent of all water on Earth. Ocean water is salty due to the presence of a chemical substance called **sodium chloride** or salt. More than 95 per cent of our oceans are still unexplored.

▲ *There is so much salt in the ocean that if you took it all out and spread it evenly over the planet's surface, it would form a layer more than 152 metres in height*

▼ *There is enough water in the oceans to fill up a cube whose edges are 1,000 kilometres in length*

Formation

This vast body of water makes Earth different from other planets in the solar system. Oceans have been present on Earth's surface for about four billion years, carving the coastline, driving the climate, and controlling life. However, 4.5 billion years ago, when Earth was born, it was a molten **inferno**. There was no place for liquid water, but the elements to create water were present deep within the newly formed planet. When the first volcanoes erupted, the gas that came out was steam. Later, molten rocks cooled and clouds formed. Consequently, it rained for centuries and filled the basins that are now our oceans.

In Real Life

Half of the oxygen present in the atmosphere is produced from ocean water.

Threat to Oceans

Despite the fact that oceans provide useful materials, they are prone to damage due to overfishing, pollution, etc. Overfishing means that people take too many fish from the water, too quickly; even faster than they can naturally be replaced. It is important to take care of aquatic ecosystems because their loss spells consequences for all living beings. The climate and weather, the air we breathe, the food we eat, all depend on oceans. Oceans are the most important source of nourishment for all the different kinds of life on Earth.

▲ *The ocean, which is home to many species of fish, has been polluted by human activities over the years*

Names of Oceans

There are five major oceans. These are the Pacific Ocean, the Indian Ocean, the Atlantic Ocean, the Arctic Ocean, and the Antarctic or Southern Ocean. The fifth ocean is the newest one considered by the International Hydrographic Organisation (IHO). However, the three major oceans of the world are the Pacific, the Atlantic, and the Indian Oceans. They are called the three great oceans. Different cultures had different names for them through the ages, but the current names were first made a long time ago—as early as 450 BCE for the Atlantic, and in the 1500s for the others.

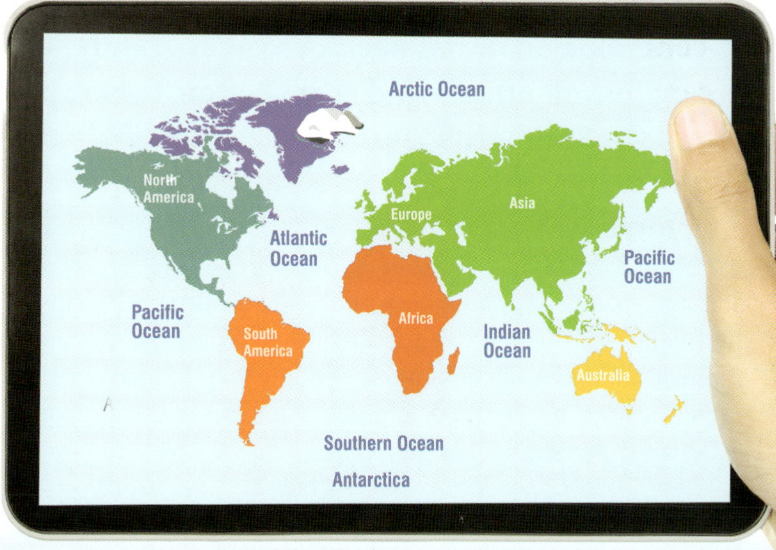

▲ *The map marks all the continents and the surrounding oceans. Actually, Earth has only one continuous ocean but it is demarcated according to landmasses*

Isn't It Amazing!

Water in the ocean is constantly and continuously moving. It is like a conveyor belt in a factory. So, the global ocean conveyor belt is the name given to the system of constantly moving and circulating water. This movement is driven by salinity and temperature.

▼ *Over the years, ports have come to be hubs of human activities*

Releasing Water

Rainfall made up only half of our present oceans. According to some scientists, the rest of the water came from space. During Earth's formative years, it was hit by thousands of **comets** which were made of ice. When the downpour stopped, Earth filled up with thousands of litres of water.

Water has always been present in Earth's mantle. It was gradually released after volcanic eruptions. In fact, water is still being released from Earth's mantle after volcanic eruptions.

The water in the oceans is attracted to the surface due to the gravitational pull of Earth. It cannot leave the planet through natural means.

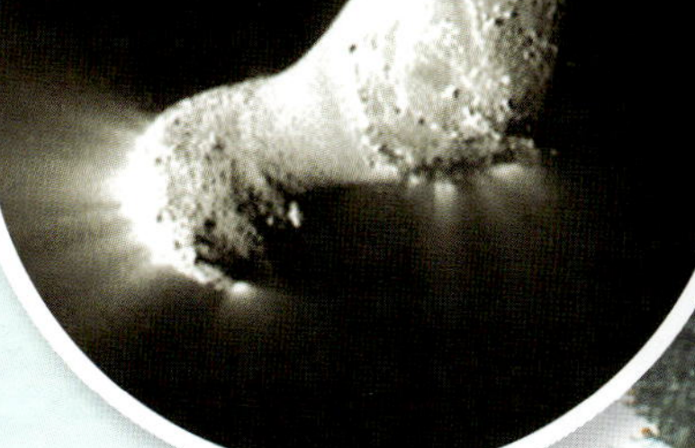

▲ *A comet is filled with dust, rock, and ice. They are informally called "dirty snowballs"*

The Five Major Oceans

There is only one true ocean on Earth. This connected body of water surrounds the seven continents. It has been demarcated into five oceans by scientists so that they can study and understand them better. These are the Pacific, Atlantic, Indian, Arctic, and the Southern Ocean.

 Why are Oceans Blue?

The water in the ocean appears blue because sunlight falls on it. Sunlight is white light. The water absorbs the red, orange, and yellow wavelengths from this white light more strongly than it does the blue wavelengths. In fact, it even reflects these blue wavelengths. That is why the ocean water and sky appear blue.

 Incredible Individuals

Film director James Cameron has made a number of trips and expeditions to the ocean deep. He has dived on the wreck of the Titanic multiple times, and is 1 of 4 people who have made their way to the depth of the Challenger Deep, the deepest point in the Mariana Trench.

The map marks the five major oceans of the world

 The Pacific Ocean

The Pacific Ocean is the biggest ocean in the world. It covers more than 30 per cent of Earth's surface. The name 'Pacific' comes from the Latin word '*pacificus*', meaning 'peaceful ocean'. However, it is not that calm and peaceful, as the Ring of Fire is located near it. There are too many volcanoes in this region, and volcanic eruptions occur quite frequently along with earthquakes.

75 per cent of the world's active volcanoes are located in the Pacific Ocean basin

The photograph shows a part of the Atlantic Ocean Road, an 8.3 kilometres road that connects the island of Averøy, Norway, with the mainland

The Atlantic Ocean

The Atlantic is the second-biggest ocean in the world and lies between the continents of America, Europe, and Africa. It is about half the size of the Pacific Ocean, covering roughly 20 per cent of Earth's surface. It was mentioned for the first time in 450 BCE as '*Atlantis thalassa*' in the book *The Histories* by Herodotus, meaning the 'Sea of Atlas'.

The Arctic Ocean

The Arctic ocean is the smallest and shallowest among the major oceans. It is located entirely within the Arctic Circle and covers around 4 per cent of Earth's surface. It occupies the region of the North Pole. Once called the Frozen Ocean, most of the Arctic Ocean is covered with ice throughout the year.

▼ The Arctic Ocean's name is derived from the Greek word 'arktikos', which means 'bear'

Arctic Ocean

Atlantic Ocean

Pacific Ocean

Indian Ocean

Southern Ocean

The Southern Ocean

The Antarctic Ocean was recently renamed as the Southern Ocean. The Atlantic, Indian, and Pacific Ocean basins merge into icy waters around Antarctica. It is the fourth-largest ocean located at the southernmost point of the planet and surrounds Antarctica. The Southern Ocean is the least explored and least understood marine region in the world.

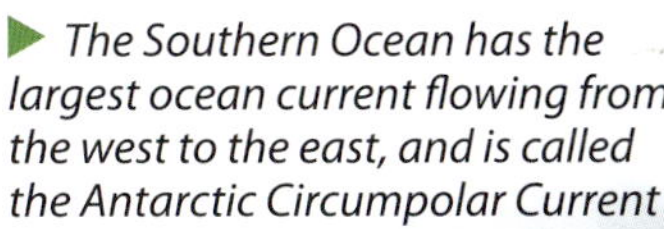

▶ The Southern Ocean has the largest ocean current flowing from the west to the east, and is called the Antarctic Circumpolar Current

▶ The deepest point in the Indian Ocean is the Sunda Trench with a depth of 7,725 metres

The Indian Ocean

The Indian Ocean covers one-fifth of the world's oceans and is the third largest after the Pacific and the Atlantic Oceans. It is geologically the youngest and physically the most complex among the three major oceans. The Indian Ocean is bounded by Australia in the East, by Africa in the West, by Asia in the North and the Atlantic Ocean in the South. It is said to be the warmest ocean of all.

Life in the Ocean

The term 'marine life' refers to all the living organisms living in Earth's waters. These include a diverse array of plants, animals, and other organisms such as **bacteria** and **archaea**. It is believed that about one million species of animals live in the ocean. Among these, 95 per cent are invertebrates such as jellyfish and shrimp.

Incredible Individuals

Marie Lebour (1876–1971) was a British marine biologist (a person who studies marine life in great detail). She researched the life cycles of different marine animals like molluscs and microplankton. She discovered 28 new species of microplankton which she described in her written works. She also conducted research on the eggs and larval stages of herrings and sprats.

Flora

An ocean is a large and deep body of water. Plants and animals are found in all different parts of the oceans except for regions where there is no sunlight. Marine plants provide food, shelter, and protection for the animals. One can find great diversity in the flora of the ocean, from **microscopic** plankton to the *Nymphaeaceae*, a giant species of the water lily plant.

There are two types of plants found in the ocean—rooted plants and floating plants. Rooted plants, as the name suggests, are plants whose roots are attached to the ocean floor. They need sunlight, so they grow near the shores and in shallow waters. Floating plants drift freely without attached roots. They grow on the water's surface.

A majority of the plants are tiny algae called phytoplankton. All these microscopic plants have a big job to do. They produce half the world's oxygen, which human beings and all other organisms require for survival. There are some bigger algae in the water called seaweed and kelp. We can even find forests in the ocean called kelp forests. They are present in cold and shallow waters. These big plants provide food and shelter to a variety of marine life.

Sea lily

Seaweeds are used in foods, fertilisers and cosmetics. They also have medicinal value

Fauna

Fish, sponges, sea anemones, and corals are some examples of marine animals. Crustaceans are invertebrates that have exoskeletons. They are also considered to be marine animals. Crabs, shrimp, and lobsters are examples of crustaceans. But marine diversity is not just limited to fish and crustaceans, even some mammals live in the water. Dolphins, porpoises, and manatees are examples of marine mammals. A common vertebrate found in the ocean is the bristlemouth fish. It is a tiny ocean fish that glows in the dark and has needle-like fangs.

Shrimp caught from the ocean

Marine animals can be divided into three groups—zooplankton, nektons, and benthos. The word plankton originates from the Greek word '*planktos*', which means 'wandering'. Zooplankton, such as jellyfish, are drifting or floating animals. They are very small. It is difficult to see these creatures with the naked eye. Nektons are free swimmers and most of the ocean animals belong to this class. Octopuses, dolphins, whales, eels, squids, and sharks are examples of nektons. Benthos are the animals that spend their entire lives at the bottom of the ocean. Lobsters, oysters, and some species of worms and snails are examples of benthos.

Some species of jellyfish are immortal. For example, the immortal jellyfish—Turritopsis dohrnii—found in the Mediterranean Sea and in the waters of Japan cannot die of old age

Nekton, such as dolphins, can live in both shallow and deep ocean waters

Coral reefs show some of the greatest bio-diversity in the world, enough so that we are still finding new species

Isn't It Amazing!

The eel uses its shock to stun its prey, and keep predators at bay. The shock of an electric eel has been known to knock a horse off its feet. Now that's shocking!

Electric eel

Importance of Oceans

Oceans cover roughly three-fourths of our planet. They are of great importance not only to human beings but to each and every living organism on Earth. Surprisingly, oceans are the world's largest producer of food. They also provide minerals, oil, and natural gas. They produce more than half the world's oxygen and absorb carbon dioxide. It is estimated that oceans produce around 50–80 per cent of the oxygen on Earth and absorb almost one third of the carbon dioxide emitted. Oceans and ocean currents maintain the global climate by absorbing heat from the Sun.

Regulating Climate

Oceans regulate the climate on Earth in many ways. The water soaks up heat and transports the warm water from the Equator to the poles, and cold water from the poles to the tropics by the movement of the ocean currents. Without these movements, the global climate would have been inhospitable.

▲ *The fishing industry worldwide employs around 200 million people*

Food Source

Oceans are a great source of protein as they contain many different species of fish that are consumed by billions of people. Algae and sea plants are used for cooking delicacies such as sushi. It is believed that there are more than 2.2 million different species of animals that live underwater. It is possible that there are species of animals yet to be discovered.

Occupation

Statistics reveal that 59.6 million people around the world are engaged in fishing and **aquaculture**. Therefore, oceans are a source of income for many. However, our oceans are facing major threats due to global warming, pollution, habitat destruction, invasive species, and a dramatic decrease in ocean fish stocks. These threats to the oceans are so extensive that more than 40 per cent of the ocean area has been severely affected and no area has been left untouched. It is therefore important for us to protect oceans.

In Real Life

There is a component in phytoplankton known as beta-carotene. It can protect the corneas of our eyes and has a lot of minerals that can improve our vision.

◀ *Phytoplankton is incredibly nutritious*

Oxygen

Though there are no trees in the oceans, algae and phytoplankton act as trees and fill the atmosphere with ample oxygen. These organisms may be microscopic, but they do a very important task.

Sources of Water

There are many types of waterbodies on Earth. They may be saltwater or freshwater. Some of these may be moving and some are contained. They can be distinguished by their shapes, sizes, and vegetation. Sources of water can be broadly divided into two categories: surface water and groundwater. Surface water is found in oceans, seas, lakes, rivers, ponds, and reservoirs. Groundwater lies under land surface that travels through and fills openings in the rocks. The rocks that store and transmit groundwater are called aquifers.

▲ *Seawater is salty, like ocean water*

Surface Water

It is a body of water that is above the ground, such as oceans, streams, rivers, lakes, wetlands, reservoirs, and creeks. Surface water moves to and from the Earth's surface through the water cycle. Precipitation and water runoff feed bodies of surface water.

There are three types of surface water: perennial, ephemeral, and man-made. Perennial type of source is permanent that persists throughout the year and is refilled with groundwater when there is little precipitation. Rivers, lakes, streams, and ponds are some examples of perennial surface water.

Ephemeral (semi-permanent) surface water exists for only part of the year. This surface water includes small creeks, lagoons, and water holes. Man-made surface water is found in artificial structures, such as dams and constructed wetlands.

The availability of surface water is naturally more than the groundwater. Therefore, it can be used for drinking, cooking, bathing, and agricultural activities. Wetlands with surface water are also important habitats for aquatic plants and wildlife.

▲ *A view of the Bay of Bengal*

Groundwater

These sources are found beneath the land surface, such as springs and wells. In the water cycle we can observe that when rain falls to the ground, some water flows along the land to the streams or lakes, other water evaporates into the atmosphere, while some is taken up by plants, and the remaining seeps into the ground. Water that seeps deep into the ground is called groundwater. It is formed as water flows through the earth's surface and passes through rock voids. Groundwater and surface water are also reservoirs that can feed into one another. Groundwater can resurface on land to replenish surface water, while surface water can seep underground to become groundwater. These are the places where springs form.

▲ *Ponds support two-thirds of all freshwater species*

Home to History

Early civilisations grew along rivers. Since the pre-historic age, rivers attracted human beings looking to settle in one place. Early settlers realised that lands along rivers were fertile enough to cultivate crops. They grew crops and stored the surplus, eliminating the need to move from place to place in search of food. They formed permanent settlements which grew into famous ancient civilisations like the Mesopotamian, Indus Valley, Nile River Valley, and the Yellow River civilisations. These further became the basis for states, nations, and empires.

◀ *The land around Rivers Tigris and Euphrates is still very fertile. As there was an abundance of water, people of the Mesopotamian Civilisation developed farming and irrigation systems*

▼ *The Nile River flooded every year in August and provided nutrients to the soil for growing crops in the dry desert land of Egypt*

▲ *The Euphrates River rises in eastern Turkey and flows through Syria and Iraq*

Tigris & Euphrates Rivers

The Tigris-Euphrates river system in southwest Asia constitutes these two rivers. They flow parallel to each other through the Middle East. Their source is in Turkey, 80 kilometres apart. They travel through northern Syria and Iraq towards the Persian Gulf. The length of River Euphrates is 2,800 kilometres. On the other hand, River Tigris is not as long as the Euphrates. It is about 1,850 kilometres long. Many civilisations have grown along the banks of these rivers. The most famous and prominent one was the Mesopotamian Civilisation which stretched from c. 5000–3500 BCE.

The Nile River

The Nile River is located in southeast Africa and has been recognised as the longest river on Earth. It rises south of the Equator and flows northward through northeast Africa to drain into the Mediterranean Sea. It has a length of about 6,650 kilometres. This mighty river has two main tributaries which meet to form the Nile. One tributary is called the White Nile, which starts in South Sudan. The other is called the Blue Nile, and that starts in Ethiopia.

This river has been very important from the perspective of human history. It provided a fertile soil for cultivation. The river therefore played an essential role in the growth of the ancient Egyptian Civilisation around c. 3100 BCE.

The Indus River

The Indus River, also known as Sindhu in Sanskrit, is a river in South Asia. It is one of the longest rivers in the world, with a length of 3,200 kilometres. It flows from Tibet, into Jammu and Kashmir, and then into Pakistan. The main tributaries of the Indus River are the Rivers Kabul and Kurram on the right bank; and the Jhelum, Chenab, Ravi, Beas, and the Sutlej Rivers on the left bank. The Indus Valley Civilisation grew around this river around c. 4000–3000 BCE. It had highly developed cities with a famous underground drainage system.

▲ The map shows the extent of the Indus Valley Civilisation at its peak, with clear markings of River Indus and its tributaries

▲ Ruins of two big cities of the Indus Valley Civilisation, Mohenjo-Daro and Harappa

▲ The length of the Yellow River is 5,463 kilometres. The area of its drainage basin is around 750,000 square kilometres

In Real Life

The Indus River displays a unique phenomenon called **tidal bore**. It is a giant wave which moves forward from the sea onto the rivers along with an unusual tide. It enters shallow and narrow inlets and can cause a lot of destruction.

The Yellow River

The Hwang Ho River in China, also known as the Yellow River, is the third largest river in Asia. It is the principal river in China and is called the 'cradle of Chinese civilisation'.

The Hwang Ho River rises on the plateau of Tibet and flows eastward, emptying into the Yellow Sea. It stretches across China for more than 4,667 kilometres. It carries yellow-coloured silt all the way, hence the name. The river and its tributaries played a vital role in shaping the Chinese civilisation.

What Are Rocks?

Rocks are made up of one or more minerals. Unlike minerals, they lack a uniform or crystalline structure. Therefore, two rocks might have different physical properties depending on how many minerals they are made of. Rocks are classified based on how they are formed. There are three major types—igneous, sedimentary, and metamorphic.

Igneous Rocks

During a volcanic reaction, magma is released from beneath the Earth's crust. This magma flows out onto the surface as lava. The magma cools down and becomes hard when it reaches Earth's surface. When it solidifies, it forms igneous rocks. Sometimes, it solidifies within Earth's crust.

The **atoms** and **molecules** of melted minerals make up magma. These rearrange themselves into mineral grains as the magma cools down. Basalt and granite are examples of igneous rocks. Large slabs of Earth's crust are made of granite. It contains the minerals quartz and feldspar. Basalt is more commonly found on the seafloor. It is a volcanic rock, rich in magnesium and iron. Volcanic rocks are formed when lava cools and solidifies on the earth's surface. Volcanic rocks are also known as 'extrusive igneous rocks' because they form from the 'extrusion,' or eruption, of lava from a volcano. Granite and basalt are also found in the lava erupted from volcanoes in Hawaii, Iceland, and some parts in the northwest of the USA.

▲ *Basaltic lava flow in Hawaii*

▶ *A basalt rock structure in Russia*

Sedimentary Rocks

Sedimentary rocks are formed mainly by the deposition of sediments carried by a river or stream to seas and oceans. Sediments are composed of minerals or organic matter. These sediments settle at the bottom of oceans and seas over a long period. Eventually, they harden into solid rocks. Some examples of sedimentary rocks are shale, limestone, and sandstone. They are also formed from eroded fragments of some other types of rocks and the remains of plants and animals.

Metamorphic Rocks

Metamorphic rocks are formed due to excessive heat and pressure on the existing igneous and sedimentary rocks. The name 'metamorphic' means 'to change form'. Therefore, as the name suggests, metamorphic rocks are transformed from sedimentary or igneous rocks because of heat, pressure, and obtrusion of fluid. The heat can come from magma or hot water from hot springs. These rocks are mainly found in the interior of Earth's crust where they are subjected to such intense pressure and heat that they transform. For example, slate is a metamorphic rock formed from shale, which is a sedimentary rock.

▲ *Cliffs made of metamorphic rocks are commonly found in Northwest Scotland and the western Isles of United Kingdom*

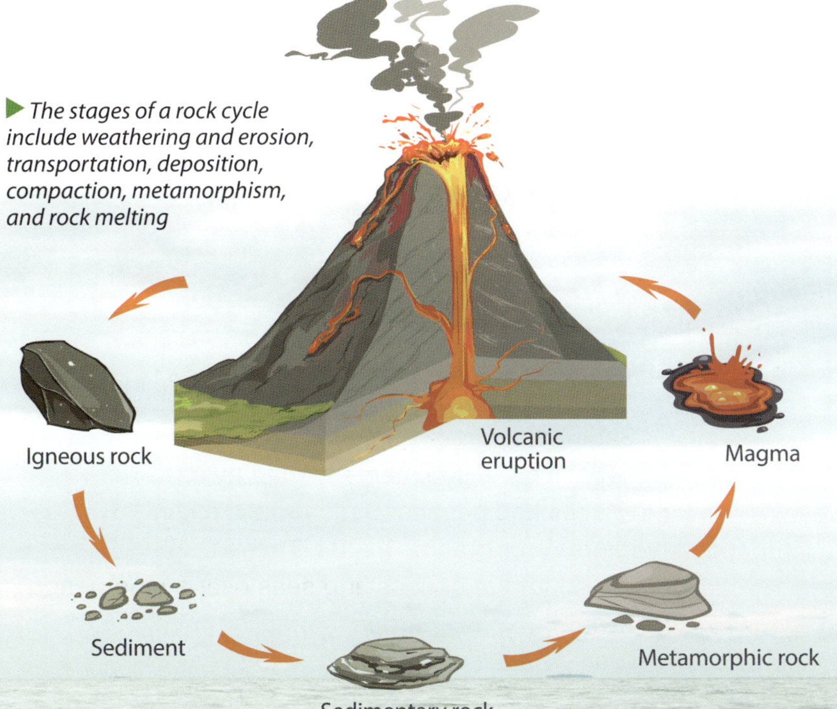

▶ *The stages of a rock cycle include weathering and erosion, transportation, deposition, compaction, metamorphism, and rock melting*

The Rock Cycle

Rocks are always changing. However, a rock takes millions of years to transform into another type of rock. The rock cycle is a geological process where all three types of rocks undergo gradual changes. For example, magma from a volcano cools down and hardens to form igneous rocks. These rocks are sometimes broken into sediments (fragments) by the forceful action of river water, harsh winds, or extreme weather. Over the years, these sediments accumulate and compress to form sedimentary rocks. Then, because of heat and pressure, the sedimentary rocks transform into metamorphic rocks. This cycle is repeated again but not necessarily in the same order. In the rock cycle, changes can occur in any order.

Incredible Individuals

▲ *A portrait of James Hutton painted in 1776*

James Hutton (1726–1797) is considered to be the father of modern geology. He came up with the theory of uniformitarianism. This gives an explanation about the processes that Earth's surface has been through in 'geologic' time. It means that Earth has changed so much over a long period that some rock types that were formed in the past cannot form today. This theory helped future geologists understand how to study rocks.

Igneous Rocks

Igneous rocks are different kinds of crystalline rocks. They are formed by the cooling and solidification of magma or lava that emerges from beneath the Earth's crust. The word 'igneous' means 'fire' as its origin is the Latin word '*ignis*'. For instance, a lot of hard rocks found in northern Canada are actually igneous rocks.

Types of Igneous Rocks

There are basically two types of igneous rocks—**extrusive** and **intrusive**. When the molten substances cool and harden on the surface of Earth, they are called extrusive. As the lava comes in contact with the cool temperature of the atmosphere, it cools down faster. This is the reason these rocks have fine grains. Rocks such as basalt, pumice, obsidian, tuff, andesite, and dacite are all examples of extrusive rocks.

▲ *Granite is an example of intrusive igneous rock*

Intrusive igneous rocks are those where the magma solidifies deep inside Earth's crust. The magma takes a long time to cool and this process goes on for millions of years. Intrusive igneous rocks vary from thin sheets to huge and irregular masses. Diorite, gabbro, granite, pegmatite, and peridotite are examples of intrusive rocks.

◀ *Pumice is an extrusive igneous rock. It has a porous surface which is the result of gas-rich frothy lava quickly solidifying*

Formation of Igneous Rocks

Due to immense pressure and heat, magma flows up to the surface of the Earth during volcanic eruptions. At this point, the temperature of the molten magma can be around 700–1300° C. Over the years, this magma crystallises and becomes a hard, igneous rock. Igneous rocks are composed of one or more minerals. If the magma cools down too quickly, then it leads to smaller-sized crystals, whereas if the cooling is slow, it forms large crystals. Therefore, the cooling process of the magma determines the size of the rocks.

In Real Life

Obsidian forms quickly with a microscopic crystal growth and is known as volcanic glass. Because obsidian has very sharp edges, it was used to make cutting tools and arrowheads. Even today, there are some surgeons who like to use scalpels made of obsidian, because they can be made 3 times sharper than diamond, and 500-1000 times sharper than a steel blade.

Texture

The extrusive and intrusive igneous rocks differ in texture. Extrusive rocks cool quickly and have **microscopic** grains. Igneous rocks are said to have an aphanitic texture. The word aphanitic is applied to igneous rocks whose surface texture is so fine-grained that it is difficult to understand their mineral composition by simply looking at them.

On the other hand, intrusive rocks that cool slowly have a texture where the grains are visible and there is significant crystal growth. This is referred to as a **phaneritic** texture.

Some igneous rocks have different textures depending on the way they are formed. Obsidian has a glassy texture because it is formed when lava cools quickly. Pumice and scoria are formed when there is an explosive volcanic eruption where the magma breaks as it reaches the surface releasing gas bubbles. Due to this, they have a **vesicular** texture. On the other hand, tuff has a pyroclastic texture as it is formed from volcanic ash. Pyroclastic rocks are formed from the broken materials ejected from the volcano.

▲ *Obsidian rocks have a glassy lustre and are harder than window glass*

Classification and Identification

Igneous rocks can be classified on the basis of their chemical composition, especially the different types of minerals present in the rocks. They are often named on the basis of the proportion of the different minerals they contain.

Geologists identify these rocks by their texture and formation. The way the magma cools down dictates the texture of the rock. For example, magma that solidifies beneath Earth's crust and forms intrusive igneous rocks will surely have a different texture from the magma that forms on Earth's crust. This magma comes into contact with ash flow as a result of volcanic ash spewing up from an eruption, flowing across and settling down to form rocks.

▼ *A grey lava field near a Hawaiian volcano. A lava field, is a large expanse of nearly flat-lying lava flows*

◀ *These basalt rock formations are called 'trolled toes'. According to Icelandic folklore, the basalt rock formations off the coast of Vik were once trolls*

Sedimentary Rocks

Sedimentary rocks are formed by the consolidation of sediments. Sediments carried by rivers, glaciers, and winds generally form them. These rocks are deposited in layers and sometimes contain fossils of plants and animals. Sedimentation is the collective name for the processes that cause these particles to settle in place.

Some sedimentary rocks are made of gravel and pebbles and are cemented by calcium carbonate

Formation of Sedimentary Rocks

Sedimentary rocks are the most commonly found rocks on Earth. They are generally formed from rocks that experience constant weathering and erosive action by wind, water, or other forces. All rocks are subject to erosion. This happens due to the force of rivers, gravity, winds, and thermal expansion of pre-existing rocks. Under their influence, these rocks are broken down into pieces. The small pieces that have broken off from the rock are then carried away over long distances. Eventually, they are deposited in layers somewhere like on the banks of a river. These layers are converted into new rocks by **compaction** (squeezing) and cementation (binding). Therefore, sedimentary rocks are formed by sediments that are collected in one place after breaking off through the process of weathering, erosion, deposition, compaction, and cementation. Sandstone, limestone, mudstone, coal, and shale are sedimentary rocks. Some sedimentary rocks may be either organic or chemical, depending on how they formed. Limestone, for example, may be created from either organic or chemical processes.

Types of Sedimentary Rocks

There are three types of sedimentary rocks—**clastic**, chemical, and organic.

Clastic

The deposits of broken rocks due to weathering form clastic sedimentary rocks by compaction and cementation. Sandstone, shale, siltstone, and breccias are examples of clastic sedimentary rocks.

Clastic sedimentary rocks are named according to the grain size of the sediment particles

Chemical

Chemical sedimentary rocks are formed when the water that the sediments are mixed with evaporates and leaves behind dissolved minerals. These rocks are quite common in arid lands or deserts due to lot of wind erosion. Examples include rock salt, flint, iron ore, and chert.

Halite is an example of a chemical sedimentary rock. It is also known as table salt

Organic

These sedimentary rocks are formed from organic debris such as roots, leaves, or animal matter. This debris has calcium minerals, which pile on the sea floor over time. Gradually, organic sedimentary rocks are formed. Examples include coal, chalk, and dolomite.

Coal is an organic sedimentary rock formed from the remains of plants squeezed deep beneath the ground over millions of years

Classification and Identification

Over the years, geologists have attempted to classify sedimentary rocks on the basis of their origin and composition. However, there is a lack of universal acceptance for any of the proposed classification systems. This is because there is a great variation in the mineral composition, texture, and other properties of sedimentary rocks.

Texture

Texture is the physical makeup of any rock, and covers the size, shape, and arrangement of the various grains or particles present in its composition. Two main textural groupings for sedimentary rocks are clastic or fragmental, and nonclastic or crystalline. Sedimentary rocks with a clastic texture are made up of grains or clasts that do not interlock, instead, they are piled together and cemented. This type of rock is porous and not very dense.

Chemical sedimentary rocks are crystalline, which means that they are made of crystals interlocked with each other. However, the same type of texture is also seen in igneous and metamorphic rocks. The textures of sedimentary rocks are not too varied as they are formed from existing rocks.

▲ *Layers of a sedimentary rock*

▼ *Mud cracks and ripple marks are features of sedimentary rocks that often contain fossils*

Metamorphic Rocks

Metamorphic rocks are those that are made up of other existing igneous and sedimentary rocks. They are usually formed due to certain changes in temperature, pressure, and the addition and subtraction of chemical components through a process called metamorphism, which means 'change in form'. These metamorphic rocks cover 12 per cent of Earth's surface.

Layers within metamorphic rocks, also known as foliation

Formation of Metamorphic Rocks

Metamorphic rocks are formed when the physical or chemical composition of the existing rocks is altered due to heat and pressure into a denser form. The movements of the tectonic plates force the pre-existing rocks to move and shift. In the process, the rocks that are buried in the interior are subjected to immense pressure and heat and are subsequently warped and deformed.

As this phenomenon goes on for many years, a plate comprising of sedimentary or igneous rocks might become subducted under another plate. The weight above can cause the rocks under it to go through the process of metamorphism. The heat from magma and the constant friction along a fault line can also bring about this type of change. In this case, rocks do not melt fully but there are definite changes in their chemical composition, texture, and even their physical structure. Pressure or temperature can even change previously metamorphosed rocks into new types. Metamorphic rocks contain the same minerals as the original rocks, but their structures are just rearranged.

Granite metamorphosed into gneiss. The process of metamorphism it goes through makes the new rock denser than the original rock and less vulnerable to erosion

Types of Metamorphism

There are three ways that metamorphism in rocks occurs: Contact, regional, and dynamic metamorphism. This can lean to either foliated or non-foliated rocks, which are separated on basis of texture and strength.

Contact, or thermal, metamorphism occurs when magma comes in contact with existing rock. This causes the temperatures to rise while the fluid from the magma seeps into the rock. It produces non-foliated rocks.

Regional metamorphism is caused by large geologic processes such as mountain-building. It occurs over a much larger area, and as the rocks are exposed to the surface, with huge pressures acting on them, they are bent and broken. It produces foliated rocks.

Dynamic metamorphism also occurs due to mountain-building, and the forces of pressure and heat cause the rocks to bent, be crushed, flattened, folded, and sheared.

Classification and Identification

Metamorphic rocks are classified by their texture, which is either foliated or non-foliated according to their parent rock. A metamorphic rock that is formed from more pressure than heat is foliated. This type of metamorphic rock has a distinct texture. It usually has a banded and layered appearance. Rocks that are formed due to tectonic movements are called nonfoliated. They do not have a layered appearance and their formation is dependent on pre-existing conditions. These rocks have a uniform look.

▼ *A cross-section of a gneiss rock with red veins*

How They Look – Texture

Foliated texture refers to the parallel arrangement of certain mineral grains, which give the rock a striped appearance. For example, gneiss, schist, slate, and phyllite have a banded appearance. These rocks develop a sheet-like structure that tells us the direction in which the pressure was applied.

Non-foliated rocks have a different texture than foliated rocks owing to the different ways these rocks are formed. Hornfels, marble, quartzite, and novaculite are good examples of non-foliated metamorphic rocks.

Isn't It Amazing!

Schist is a metamorphic rock that originates from slate, another metamorphic rock. In Australia, houses that were built using schist in the 1800s are still standing today. Presently, schist is used as a decorative stone as well as for jewellery.

▲ *Here, schist stones have been piled up to form a wall*

▼ *Metamorphic rock formations in Montana, USA*

Soil

Soil is the uppermost layer of Earth's crust where plants can grow. It is a mixture of organic matter such as decaying plants and animals, and fragmented rocks and minerals. This layer is not as solid as rock but has small spaces known as pores that are capable of holding air and water. It looks like dirt. Are soil and dirt the same? No, dirt is soil that has lost its ability to support life.

Formation of Soil

Soil takes a long time to form. It is believed that it can take thousands of years to form one inch of soil. However, other factors influence its formation. All living organisms such as plants, fungi, animals, and bacteria help in the process. Other factors like the **topography** of a place and the climate are also vital in soil formation. One important factor, of course, is the parent material. It refers to the kind of rocks and minerals that slowly disintegrate to form soil. Soil is vital for the sustenance of life on Earth.

In Real Life

You can grow different plants based on the soil that is available. Gardeners and professional landscapers choose their plants according to the soil. For example, if sandy soil is abundantly available, a gardener would plant broom shrubs that do not need more water and can grow well in sunny places. In peaty soil, a gardener might grow the heather shrub that grows low on the ground and spreads across the available area.

Layers of soil

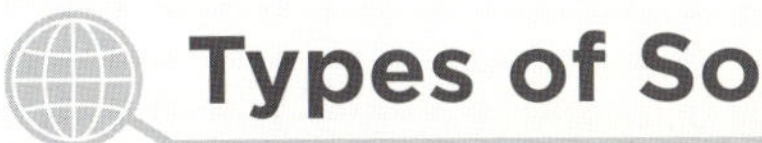

Types of Soil

There are different types of soil: sandy, clay, loamy, silt, peat, and chalk. Soil is differentiated on the basis of its composition.

Sandy soil: This type of soil is light and not very rich in nutrients. The particles are fine and hard. Sandy soil does not bind well and cannot hold water for a long time. It is formed mainly from the weathering of bedrock such as shale, limestone, granite, and quartz.

Clay soil: It is smoother than sandy soil. The particles of this type of soil are fine and bind well. It is a wet and sticky soil. While wet, it can be moulded into different shapes. However, dry clay soil is quite hard. Clay soil, unlike sandy soil, does not drain well.

Chalky soil: It is not acidic, always alkaline. It is a strong soil containing chalk or limestone. It can hold little water and gets dry quickly. This type of soil is fertile, but not all the nutrients are present in it.

Silt soil: It has a smooth texture and holds water better than sandy soil. This type is good for cultivation. Silt soil is formed when sediments carried by water or ice get deposited. Sometimes, the silt that gets deposited becomes silt stone over time.

Loamy soil: It is a mixture of sand, silt, and clay particles. It can retain water. This type of soil is ideal for cultivation as it is full of nutrients from decomposed material.

Weathering, Erosion, and Conservation

Weathering refers to the breaking down of rocks and minerals on Earth. The agents of weathering are water, ice, acids, **salts**, plants, animals, and changes in temperature. The process of weathering takes place over millions of years to produce a thin layer of soil.

Types of Weathering

There are three types of weathering—physical, chemical, and biological. Physical weathering occurs due to the effect of temperature changes on rocks, which causes them to break. Sometimes, water can help this process along. It takes place mainly in mountainous areas and deserts where there is little soil and scarce vegetation to cover the soil.

Chemical weathering is caused by rain water that reacts with mineral grains in rocks to form new minerals. The water, in this case, is acidic in nature. This phenomenon can only take place when there is sufficient water. It happens at higher temperatures. Therefore, warm and humid climates suit it the most.

Plants and animals cause biological weathering. They produce acid-forming chemicals that cause weathering and, in the process, break down rocks.

▲ *The effects of physical weathering created this great natural structure*

Erosion

After the rocks are broken down, they form soil. Erosion happens on this soil. It is the process by which weathered soil, rocks, and other materials on Earth's surface are dispersed by the actions of gravity, water, ice, and wind. In short, it is the transportation of weathered rocks.

Erosion happens when streams and rivers wear down rocks and carve caves and crevices. The minerals behave like sandpaper and rub off the rocks. As the smaller pieces move downstream, they scrape larger rocks and soften them. Winds in the desert send tiny particles flying through the air. These particles can change the shape of rocks and landscapes. Glaciers can rip away fragments of rocks when they move. They can even create a valley.

Conservation of Soil

Earth is a beautiful planet with natural resources like air, water, soil, minerals, fuels, plants, and animals. Conservation is the practice of caring for these resources for the benefit of all living beings. Soil is one of the natural resources. It is instrumental in the production of food. A certain quality of soil is required for the cultivation of crops to feed human beings and animals. Therefore, it is vital to conserve soil for the well-being of all living things.

Sowing the same crop repeatedly in one place can reduce the nutrients in the soil. This is called 'monoculture'. To avoid this, the conservation method known as 'contour-strip cropping' can be applied. Plants such as wheat, corn, and clover should be planted in alternate strips across the path of prevailing winds. Different crops have different types of leaves and root systems that slow down erosion. Another method of soil conservation is 'selective harvesting'. It is the practice of removing certain individual trees or small groups of trees while leaving others rooted in place to anchor the soil. It is important to conserve soil to save it from harsh weather and erosion. Proper conservation yields more nutritious soil.

In Real Life

Soil is alive but it can also die. Due to the allure of short-term gains, excessive use of chemical pesticides and fertilizers causes long-term damage to soil. The best example of this phenomenon is the American Dust Bowl in the 1930s. 100 million acres of fertile grasslands were destroyed at that time because of poor land management.

◀ *Soil blown by dust bowl winds piled up in large drifts on a Kansas farm*

▼ *Bank erosion is a natural process without which rivers would meander or change their course*

▶ *Contour-strip cropping is used when there is no alternative method for soil conservation while farming on a slope*

Minerals

Minerals are solid, inorganic crystalline substances. They are formed naturally, and each mineral has a specific chemical composition. Minerals that are made up of one element are called native minerals, for example, gold, silver, copper etc. However, when a mineral has more than one element, it is a **compound** mineral. Being made up of silicon and oxygen, quartz is an example of compound mineral.

▲ *Most minerals have a specific arrangement of atoms and ions that defines the characteristics of their crystal forms*

Occurrence and Formation

Minerals are formed in all kinds of environments and under a varied range of chemical and physical conditions. There are four methods of mineral formation. They are igneous where minerals crystallise from a rocky or molten magma. They are sedimentary when the minerals are the result of sedimentation. They are metamorphic when new minerals form from already existing ones. This metamorphosis occurs due to the change in temperature and pressure (or both) on some existing rocks. Minerals can also be formed from a hydrothermal process, wherein minerals are left over from hot-mineral laden water.

▲ *Green Torbernite Mineral in its pure form*

Structure

A mineral is pure and has a uniform structure. Solids that occur naturally are considered true minerals. Rocks consist of minerals that are grouped together. The combination of minerals determines the type of rock formed. Minerals and rocks affect landform development and form natural resources such as gold, tin, iron, marble, and granite. Minerals can be written as a single chemical formula because they are pure. However, there are some impurities too. There are over 3,000 known minerals, and the list is still growing. Some commonly found minerals are quartz, mica, olivine; while emerald lies at the more precious end of the range.

Incredible Individuals

Friedrich Mohs (1773–1839) was a German mineralogist and geologist. In 1812, he developed a decimal scale to measure the hardness of minerals. This scale is used till date. The Mohs hardness scale takes the rough measure of the resistance put up by a mineral when there is scratching or abrasive action on its smooth surface. There are numerical values assigned to the hardness. For example, talc has the value of 1 because it can easily be scratched even by a person's fingernail. Diamond has the value of 10 because you need to use a glass cutter to make changes to it.

Physical Properties of Minerals

The physical properties of minerals are determined by their chemical structure. To identify minerals, we must first identify their physical properties clearly, which are as follows—(i) colour of the mineral, (ii) streak, which is the colour of the mineral powder, (iii) lustre or the appearance of minerals in reflected light, (iv) cleavage or the tendency of a mineral to break along smooth planes parallel to the zones of weak bonding, (v) fractures or the way a mineral breaks along curved surfaces without a definite shape, (vi) hardness or the mineral's resistance to **abrasion**, (vii) specific gravity or the ratio between the weight of a substance and the weight of an equal volume of water, (viii) tenacity or the resistance to breaking, crushing, bending and tearing.

▲ *Streak is determined using a streak plate*

Grouping Minerals

Mineralogists group minerals according to their chemical composition. However, the most commonly followed grouping system is the one devised by Professor James Dana of Yale University, way back in 1848. Dana system divides minerals into the following basic classes— silicates, oxides, sulphides, sulphates, halides, native elements, carbonates, phosphates, and mineraloids.

Oxides

Oxides are chemical compounds which are made up of oxygen and other chemical elements. A majority of Earth's crust consists of oxides. They are formed when elements are oxidised by air, i.e. when oxygen in the air reacts with the element. Water is the oxide of hydrogen. Nitrous oxide is an oxide of nitrogen, commonly known as laughing gas. Many metals form oxides, for instance, those of iron, aluminium, tin and zinc. These are also important as **ores**.

▲ *Magnesium covered with a black oxide layer*

Silicates

Silicates are the largest group of minerals found on Earth. They are found in abundance and comprise silicon and oxygen. There are more silicates than all other minerals put together. They also have in them one or more metals like aluminium, calcium, potassium, sodium, barium, iron, and magnesium. Almost all of the igneous-rock-forming minerals are silicates. Some of the common silicate minerals are feldspar, quartz, granite, asbestos, hornblende, clay, and mica. All these minerals have a chemical formula. The symbol for the chemical element of silicon is Si.

▲ *A specimen of a feldspar rock*

In Real Life

Silicon's melting point is 1,414° C, and its boiling point is 3,265° C.

Sulphides

Sulphide is a member of a group of compounds of sulphur with one or more metals. Most sulphides are structurally simple and have many of the properties of metals such as metallic lustre and electrical conductivity. They have a striking colour, less hardness, and a high specific gravity. Sulphides are generally opaque, dense, and have distinct streaks. They are an important group of minerals, which include the majority of ore minerals like iron, copper, nickel, lead, cobalt, zinc, and silver.

Sulphates

Sulphate is a salt of sulphuric acid. A sulphate ion is a group of atoms with the chemical formula SO4-2. They are formed by replacing hydrogen with a metal, for example sodium. When the hydrogen atoms are replaced, then it becomes a normal sulphate. When one hydrogen atom is replaced, it is called a bisulphate. Mostly, metal sulphates are easily soluble in water, whereas calcium and mercuric sulphates are only slightly soluble. Sulphates can be found quite in abundance in nature, for example, barium sulphate occurs as barite and calcium sulphate occurs as gypsum and alabaster.

Incredible Individuals

Jacob Berzelius, a Swedish chemist, discovered silicon in 1824. He discovered it by heating chips of potassium in a silica container and then carefully washing away the residual by-products. He is considered to be one of the founders of modern chemistry. He also developed chemical symbols and noted the atomic weights of many elements. He discovered several new elements, such as Cerium (Ce) and Thorium (Th).

▲ *Magnesium sulphate is also known as Epsom salt*

▼ *Mica and feldspar are two examples of silicate minerals*

Native Elements

Native elements are chemical elements that may occur in nature uncombined with any other elements. They are composed of atoms from a single element. Elements that are present in the atmosphere are not included in this category.

Copper

Classification

Native elements are divided into three categories. The first category is metals. It includes platinum, iridium, osmium, iron, zinc, tin, gold, silver, copper, mercury, lead, and chromium. The second category consists of semi-metals like bismuth, antimony, arsenic, tellurium, and selenium. The third category is that of non-metals such as sulphur and carbon.

Native elements can form under contrasting conditions and in all types of rocks. Some metals and non-metals are found in abundance on Earth. They are of great commercial importance. For example, gold, a metal, and carbon, a non-metal, are of great importance.

In Real Life

Our bodies contain about 0.2 milligrams of gold, most of it is in our blood.

Gold is a metal that can resist natural weathering processes

The Mineral Gold

Gold is the most well-known mineral on Earth. It is known for its value from the earliest times. There are traces of silver in the natural form of gold. Sometimes, traces of copper and iron are also found along with gold. Mineral gold is bright yellow, but if the silver content is more, then it appears whiter. Gold is mined from ores like iron-stained rocks or huge white quartz. The colour of the ore is generally brown. It is crushed and then the gold is extracted from it.

The Mineral Silver

Silver has a white, metallic lustre known for its decorative beauty and electrical conductivity. Similar to gold, silver is also a valuable metal. It is found widely in nature, but when compared with other metals, the amount seems to be quite small. It constitutes 0.05 part per million of Earth's crust. Silver is the best reflector of visible light and is used to make mirrors. It has the tendency to stain or become discoloured. It becomes dark grey when it meets air, and needs to be polished intermittently.

The world's top producers of silver presently are Mexico, Peru and China

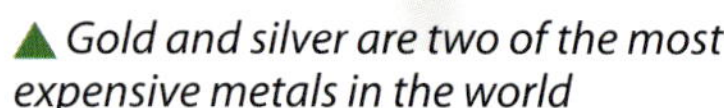

Gold and silver are two of the most expensive metals in the world

Isn't It Amazing!

There is enough gold in Earth's core to coat the planet's surface to a depth of 1.5 feet.

Other Mineral Groups

Other mineral groups are carbonates, phosphates, halides, and mineraloids.

Carbonate

Calcium carbonate

A carbonate is a chemical compound that is made of carbon and oxygen. It produces carbon dioxide, water, and a chemical salt when added to an acid. It is found naturally on Earth. Widely used carbonates are sodium carbonate (Na_2CO_3), sodium bicarbonate ($NaHCO_3$) and potassium carbonate (K_2CO_3).

A very common carbonate is ammonium carbonate, popularly known as smelling salt. Calcium carbonate is found in the shells of animals and in rocks like Iceland spar, limestone, and marble. It is also used in the production of lime or calcium oxide. Ammonium, potassium, and sodium carbonates are easily soluble in water. Carbonates give off carbon dioxide when treated with dilute acids like hydrochloric acid.

Water pools capped with calcium carbonate in Pamukkale in Central Turkey

A phosphate mine processing mill

Phosphates

Rock phosphate

A phosphate is a salt of phosphoric acid. The most important phosphate is calcium phosphate, which is available in abundance on Earth. It is not very good for plants, as it is slightly soluble in water. When it is treated with sulphuric acid, a superphosphate of lime is formed.

Phosphate rocks are the main source of phosphorous, which is not found freely in nature. Phosphate rocks are found in North Africa, Russia, and in several places in the United States.

Mineraloids

Opal

A mineraloid is a naturally occurring, inorganic solid that does not present crystallinity. It looks like a mineral but does not have the required atomic structure to be defined as a mineral. Some even do not have the necessary chemical compositions.

Some familiar materials can be described as mineraloids. For example, opal is an hydrated silica with the chemical composition $SiO_2.nH_2O$. The 'n' in its formula indicates that the amount of water is variable. Therefore, opal is a mineraloid. Another example of a mineraloid is obsidian.

Crystals of Himalayan halite salt

A rare opal boulder in Coober Pedy, Australia

Halides

A piece of fluorite

Halides are compounds comprising halogens. A halogen element is one of the five non-metallic elements that make up a mineral group. Halides produce salt. They are basically formed by combining a metal with one of the five halogen elements such as chlorine, bromine, fluorine, iodine, and astatine. As most of these compounds are soluble in water, they occur under special conditions. Halite (NaCl) or rock salt is an exception. It is very common and found in huge deposits all over the world. It is a mineral that has many uses, including the production of table salt.

What Is a Volcano?

The word 'volcano' is derived from the word 'Vulcan', the Roman god of fire. A volcano is a vent in Earth's crust through which molten lava, gases, ash, and rock fragments escape into the air. Hot gases and melted rock or **magma** are ejected from deep within Earth's crust that find their way up. This material within Earth's crust can flow slowly out of a **fissure** in the ground, but can also erupt suddenly into the air.

Volcanic Eruptions

Volcanoes have played an important role in the formation of the oceans and continents. Volcanic activity is the process by which the materials from Earth's interior reach the surface. Eruptions mainly occur when excessive pressure builds up within Earth's innermost layer, and most volcanoes have a volcanic **crater** at the top through which these materials come out.

▲ *Iceland's Kerid crater lake stands out because it is surrounded by rare red volcanic earth*

Strong volcanic eruptions throw bits of magma into the air, which cool into tiny pieces of rock called volcanic ash. Winds carry this volcanic ash thousands of kilometres away. Steam and poisonous gases also escape from volcanoes. Sometimes these gases are mixed with ash and other hot **debris**. This mixture travels outward in destructive fiery clouds called **pyroclastic flows**. Millions of years ago, hot gases like carbon dioxide streamed from within and out through countless volcanoes to form the atmosphere around the Earth and oceans.

▶ *An eruption creates a vertical column of ash and smoke. Depending on the strength of the eruption, the plume can go from under 100 metres all the way to over 40 kilometers high*

In Real Life

▲ *Gases emerging from Mount Aso in Japan*

Japan experiences frequent volcanic eruptions. There are around 60 active volcanoes in the country, and there have been three major volcanic eruptions since 1980. The reason for all of this stressful activity is the movement of tectonic plates near Japan.

How Are Volcanoes Formed?

Volcanoes do not just form anywhere on our planet. There are certain locations where these openings are formed. The main three places where they originate are divergent plate boundaries, convergent plate boundaries, and hot spots.

◀ *Lava usually emerges from the Earth at 700°C–1250°C*

Divergent Plate Boundaries

Earth's crust is made up of massive, irregularly shaped slabs of solid rock known as tectonic plates. These plates are constantly moving but in slow motion, and most volcanoes lie along the boundaries between the plates. Some volcanoes form when two plates pull apart and form divergent plate boundaries. The molten rock rises between these two plates, causing fissure eruptions in which lava flows out over the ground without any explosions. This type of volcano is found along the Mid-Atlantic **Ridge**. Volcanoes in the northern part of this ridge formed the island country of Iceland.

Convergent Plate Boundaries

Sometimes oceanic and continental plates collide. The oceanic plates are thinner and more dense, and are sometimes forced beneath the continental plates, which are thicker and less dense. This is how convergent plate boundaries form. The process by which the plate is forced down is called **subduction**.

When this happens, violent eruptions take place and force magma to rise to the surface, creating explosive volcanoes. These types of volcanoes are found around the edges of the Pacific Ocean.

▼ *Molten rock is known as magma while underground, becoming lava when it reaches the surface. Both produce igneous rock after cooling down*

Hot Spots

A small number of volcanoes form at the 'hot spots' in the Earth's crust where molten rock rises from deep below the crust. Hot spots occur in mantle plumes that are unusually hot areas and not at the boundaries of tectonic plates. The volcanoes of Hawaii are the best examples of hot-spot volcanoes.

▶ *A mud pit formed in a volcanic hot spot*

Parts of a Volcano

In this world, we can see many types and shapes of volcanoes. But all of them basically have the same parts and layers. These are either located in the mountainous regions or within the Earth. Scientists have discovered volcanoes on the moon and other planets as well, but even those have the same basic parts. These parts are magma chamber, dike, lava, main vent, side vent, throat, crater, pyroclastic flow, and ash cloud.

Lava

Lava is molten rock or semi-fluid rock that erupts out of a volcano. When lava erupts from a vent it comes out at a temperature of between 700°–1250° C. As it flows down the volcano, it is called lava flow. It eventually cools and hardens when it comes in contact with air. This solidified form is also referred to as lava. This should not be confused with magma, which is hot liquid rock that is found just below the surface of Earth.

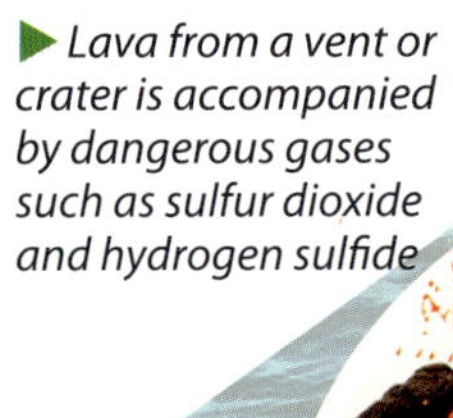

Lava from a vent or crater is accompanied by dangerous gases such as sulfur dioxide and hydrogen sulfide

Magma Chamber

It is a large pool of molten rock found underneath Earth's crust, and are usually found 1-10 km below the surface. Magma is formed when the temperatures in Earth's crust or mantle increase and air pressures decrease. It is composed of molten rock, dissolved gases, and crystals. The molten rock itself is made of oxygen, aluminium, silicon, magnesium, calcium, iron, etc. As the magma (which is a liquid) cools, it forms crystals of different metals. Eventually, it solidifies and becomes igneous or magmatic rock. When a volcano erupts, or a deep crack occurs in the crust, the magma rises and overflows from the magma chamber.

If lava cools at a faster rate, the rock it forms will have a fin grain and a glassy appearanc it cools slower, it will have larg grains and be more coarse

Volcanic Ash and Ash Cloud

Volcanic ash is formed during explosive eruptions when gases in magma expand and escape violently into the atmosphere. Hence, an ash cloud is created when this ash is propelled into the air. Volcanic ash consists of particles of rock, minerals, and glass.

Crater

The crater is a bowl-shaped opening at the top of a volcano. The outward explosion of rocks and other materials during a volcanic eruption causes the formation of craters.

Ash cloud is seen rising from the crater

Parts of a volcano

Conduit

This is the main passage for the magma to reach the surface. It is like a pipe or tube that is located at the main vent where all materials pile up. These materials move from the magma chamber to the surface from the conduit tube. A large volcano could have a network of conduit pipes where materials travel. After an explosion, some lava cools at the mouth of the conduit pipe, creating a plug.

Dike

Magma inside the mountain exerts a lot of pressure on the crust as well as on the volcano itself. The magma pushes its way through small cracks in the crust and ultimately reaches the surface, creating dikes in the process. Therefore, a dike is an **intrusion** of magma that cuts through the existing rocks and then solidifies. A volcano can have hundreds of dikes.

Main Vent

It is the weak point on Earth's crust where hot magma rises from the magma chamber and reaches the surface. It is an opening through which lava, gases, ash, and other explosive materials come out.

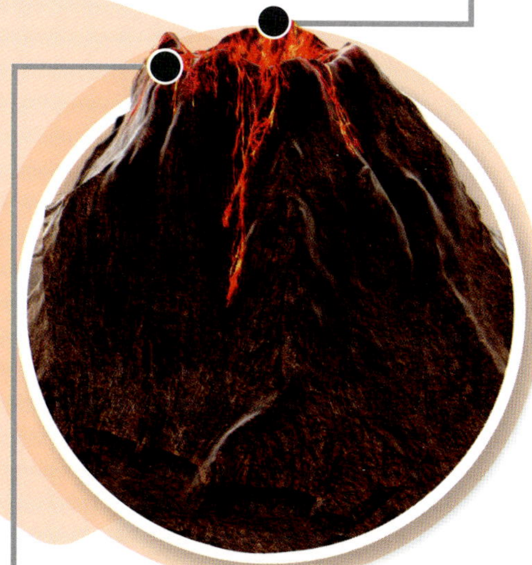

Throat

The uppermost section of the main vent is called the throat of a volcano. It is from here that the lava is ejected.

Side Vent

A side vent is an opening in the side of a volcano through which volcanic materials like lava, gases, and pyroclastic debris come out.

Categories & Types of Volcanoes

Volcanoes are a magnificent phenomenon of nature and play an important role in shaping our planet's geography. There are three categories of volcanoes: active, dormant, and extinct. This categorisation is based on the recent activity of a particular volcano and its frequency of eruption, size, and potential impact. They can be further differentiated by types as well.

An erupting volcano

Extinct Volcanoes

An extinct volcano is one that has erupted thousands of years ago and no one expects another eruption presently. One such volcano is located on Hawaii's Big Island and its name is Mount Kohala. This volcano last erupted approximately 60,000 years ago.

Volcanoes that have got cut off from their magma supply are termed as extinct. The Hawaiian-Emperor seamount chain in the Pacific Ocean, the Shiprock volcano in Navajo Nation territory in New Mexico, and the volcanoes of Edinburgh Castle in Scotland are examples of extinct volcanoes.

This is Mount Kirkjufell, an extinct volcano in Iceland.

Dormant Volcanoes

A **dormant** volcano is one that has not erupted for a while but can erupt in the near future. Sometimes, there is very little difference between a dormant volcano and an active one. This is because even though a volcano can remain dormant for hundreds of years, there is always a chance that it can erupt at some point in the future.

The Fourpeaked Mountain in Alaska is a dormant volcano.

Active Volcanoes

An active volcano is one that has erupted in the recent past, and can erupt in the future. One of the best examples of active volcanoes is Mount Kilauea, which has been erupting continuously from 1983 to the present day.

An active volcano can be quite dangerous and destructive. However, when passive, the soil it enriches can become fertile and perfect for agricultural settlements. According to scientists, when a volcano shows signs of unrest (gas emission and unusual tectonic activity) it can be called active as it is about to erupt.

Mount Kilauea erupted on 3 May 2018

Incredible Individuals

People who study or research volcanoes are referred to as volcanologists. Maurice and Catherine 'Katia' Krafft were volcanologists who dedicated their lives to the study of volcanoes. They were attracted by erupting volcanoes and saved up money to travel to Mount Stromboli to take pictures. They visited hundreds of volcanoes around the world in their lifetimes. They were nicknamed 'Volcano Devils'. Unfortunately, they lost their lives when they were hit by pyroclastic flow of the Mount Unzen volcano in Japan.

History's Deadliest Eruptions

Though Mount Vesuvius is remembered by many, perhaps the deadliest volcanic eruptions occurred in Indonesia. During the 1815 eruption of Mount Tambora, 92,000 people were killed. Again, in 1883 when Mount Krakatau erupted, it killed 36,417 people.

The Tambora Eruption

Mount Tambora is 13,000 feet in height. When this massive volcano erupted it released gas, dust, and rocks that cloaked the entire island of Sumbawa and went further still. The ash and lava that spilled down the mountains destroyed entire forests. The nearby Java Sea experienced a tsunami, that went on to kill 10,000 people.

As a result of this volcanic eruption and tsunami in Indonesia, many parts of Europe and North America suffered a famine. Volcanic ash that settled on Earth's surface after the eruption caused the destruction of the vegetation and crops.

Its huge caldera is 6 km in diameter, and 1,100 metres deep

The Krakatau Eruption

Krakatau erupted in 1883. It is an island volcano that lies between the islands of Sumatra and Java, a hotbed of volcanic activity.

When Krakatau erupted, the explosions were heard 4,653 kilometres away, all the way across the Indian ocean on Rodriguez Island. This distance is nearly one-thirteenth of the Earth's circumference. The surrounding waters rose 40 metres in height. These fearsome sea waves killed thousands.

The Krakatau explosion was the loudest sound ever recorded by humanity

Mount Thera

The eruption of Mount Thera in Greece was a catastrophic volcanic eruption that took place approximately 3,500 years ago. It devastated the island of Thera, which got buried under a thick layer of ash and pumice after the eruption. Then, a devastating tsunami occurred on the north coast of Crete. The explosion was equivalent to the blast of 40 atomic bombs. It blew out the interior of the island and forever changed its **topography**.

The caldera of Thera and the town formed near it

Huaynaputina Volcano

Huaynaputina is a young volcano located in southern Peru. It was the site of a catastrophic eruption—which was preceded by several earthquakes—in 1600 CE. The eruption is notable, not only because the summit elevation of this volcano is 4,800 metres, but also because it is the only major explosive eruption in the history of the Central Andes. It also had a great impact on the global climate. It has been recorded as the largest volcanic explosion.

While the actual eruption only lasted 12-19 hours, the air was not clear of ash from February to April

Mount Pelee

Mount Pelee is an active volcano on the Caribbean island of Martinique in West Indies. It erupted on May 1902, killing approximately 30,000 people. Casualties were mainly caused by the fast-moving **pyroclastic flows**.

The summers following the Mount Pelee eruption were the coldest compared to the previous five centuries

Tremors and Terrors

Like volcanic eruptions, earthquakes are also natural disasters. An earthquake is a sudden shaking of the ground due to movements of the large masses of rocks beneath Earth's crust. Earthquakes are constantly happening around the world. However, we do not feel the majority of them because they are too small to be significant. Like volcanic eruptions, earthquakes cause great damage to life and property.

Causes

As we know, Earth's crust is made up of large rock slabs or rock masses called tectonic plates that move slowly but constantly. These tectonic plates collide, diverge or move against the edges. Most of the time, the tectonic plates brush against each other without causing any damage. Over time, this movement causes great pressure to build up.

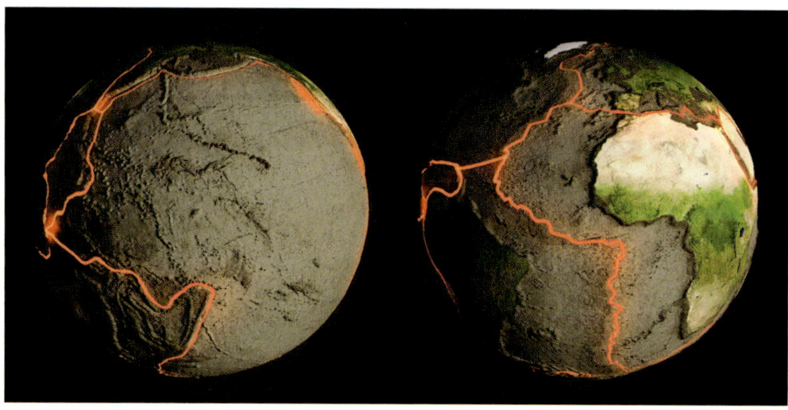

▲ *The orange lines are Earth's fault lines between the tectonic plates*

When the pressure becomes immense, the tectonic plates suddenly shift along a crack in the crust, called a **fault**. This sudden release of energy causes seismic waves that make the ground shake. Two blocks of rock or two tectonic plates rub against each other, and after a while the rocks break because of the pressure. When these rocks break, an earthquake occurs. Before and after a large earthquake, there are smaller earthquakes: the ones that happen before the main earthquake are called foreshocks, and the ones that happen when the earthquake is over are called aftershocks.

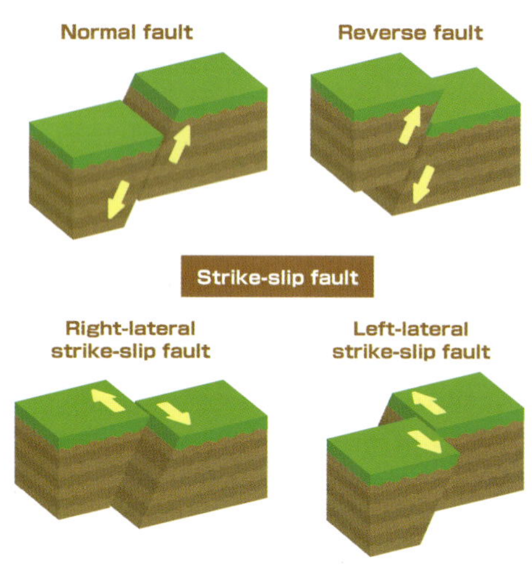

▲ *Strike-slip faults are mostly vertical, while normal and reverse faults are often at an angle to the surface of the Earth*

Types of Faults

Faults are classified into three types—normal, reverse, and strike-slip. Normal and reverse faults are also known as dip-slip faults. In a normal fault, the cracks form as one rock mass slides downwards and moves away from the other rock mass. They occur along divergent plate boundaries. In a reverse fault, the cracks form as one rock mass collides with another rock mass. They occur along convergent plate boundaries but also along areas where the plates are folding up. Here, one rock mass slides beneath another and pushes it upwards. In a strike-slip fault, the cracks form as two rock masses slide along each other's edges.

Where Do Most Earthquakes Occur?

Most earthquakes occur along a fault line or the tectonic plate edges like those in the Pacific Ring of Fire, Alpide Belt, and the mid-Atlantic Belt. An earthquake can strike in any part of Earth, but it occurs mainly in the above-mentioned areas.

Studying Earthquakes

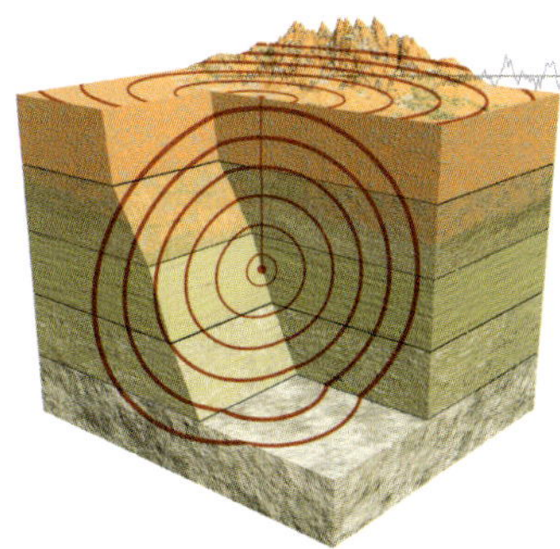

▲ A cross-section showing the epicentre and impact of an earthquake

Earthquakes are also known as tremors. In ancient times, people had all sorts of ideas about earthquakes. The earthquake of 1750 in England and a tsunami in 1755 moved scientists to learn more about these natural phenomena. This event marks the beginning of the modern era of seismology, encouraging numerous studies into the causes, locations, and timing of earthquakes.

Seismic Waves and Seismology

The shifting of rocks during an earthquake causes **vibrations**. These are seismic waves that travel within Earth's surface. They are also caused due to volcanic activity, magma movement, and landslides. Seismic waves are broadly divided into body waves, i.e., waves that travel within the body of Earth, and surface waves that can only move along the surface of the Earth like ripples on water.

Seismology is the study of earthquakes and seismic waves that move through and around Earth. A seismologist is a scientist who studies earthquakes and seismic waves. They are recorded with something called a seismograph. This instrument detects and measures the movement of the Earth.

▲ A record produced by a seismograph on a display screen or paper is called a seismogram

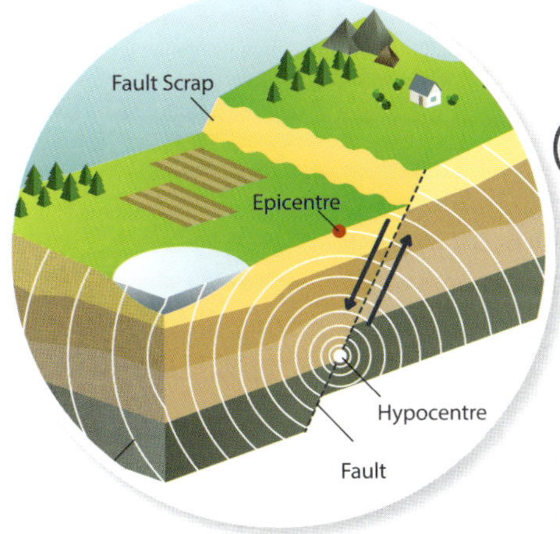

▲ Seismic waves spread out radically with decreasing intensity from both the epicentre and the hypocentre

Hypocentre and Epicentre

The **hypocentre**, or focus, is the point where an earthquake begins. It is always found below the surface of Earth. The place directly above this on the surface is called the **epicentre**. Earthquakes are the strongest at this point on the surface, and the destruction caused here is directly proportional to the size and depth of the rupture of rocks.

So, the epicentre and hypocentre both mark the origin of the earthquake. However, the epicentre occurs on Earth's surface while the hypocentre occurs beneath it. With the help of a seismogram, experts can tell how far the earthquake will spread and how strong it is.

The Richter Scale

The strength of an earthquake can be determined either by the destruction it has caused, or by careful calculations made using the readings on a scale called the Richter Scale. This scale is used to measure an earthquake's **magnitude** (size or extent). It was devised in 1935 by seismologists from USA named Charles Francis Richter and Beno Gutenberg.

Incredible Individuals

Charles Francis Richter (1900–85) and Beno Gutenberg developed the Richter scale to measure the intensity of moderate earthquakes. Charles Richter was born near a town called Hamilton in Ohio, USA. After attending the University of Southern California, he graduated from Stanford University in 1920 and received a doctorate in physics from the California Institute of Technology in 1928. Richter worked at the Seismological Laboratory of the Carnegie Institution of Washington from 1927–36. The Richter scale was published in 1935 and soon became the standard measure of earthquake intensity.

▲ Charles Francis Richter

The Aftermath

The effects of earthquakes are massive. They destroy life and property, and in the process, affect the mental and physical health of those who have encountered it. Some earthquakes are so devastating that they can erase entire towns and cities from the map. The type of impact mostly depends upon where they are occurring, whether in an urban or a rural one. It also depends on the population of that place, and of course, the ability of the infrastructure to bear the tremors.

Fires

Earthquakes can cause great damage to buildings, bridges, pipelines, railways, embankments, dams, and other structures. Changes at the Earth's surface due to earthquakes include changes in the flow of groundwater, landslides, and mudflows. Fires are also associated with earthquakes because fuel pipelines rupture, and electrical lines are damaged when the ground shakes.

The 1906 San Francisco earthquake is the best example of the destruction an earthquake can cause. It caused large fires in the downtown area of the city. These fires burned for three days and caused some of the greatest property losses ever seen. Fires, as a by-product of earthquakes, also damaged life and property in the Tokyo earthquake of 1923.

▲ *San Francisco burning after the 1906 earthquake*

Tsunami

Underwater earthquakes can cause gigantic waves to move towards the shore. This is called a tsunami. 'Tsunami' is a Japanese word for 'harbour wave'. During a tsunami, a series of waves in the ocean force the water to first pull back from the shore, and then move towards the shore, sometimes reaching heights of hundred feet as they travel. These waves can travel very fast and by the time they reach the shore, they gain a lot of speed. Tsunamis have the ability to destroy coastal settlements.

Liquefaction

During an earthquake, the ground shakes so much that it can loosen rock and unconsolidated materials to such an extent that they can transform a solid rock into a mass that flows like a liquid. This process is called liquefaction.

Isn't It Amazing!

The San Francisco earthquake was the first natural disaster to be photographed.

Safety Measures

Volcanic eruptions and earthquakes are natural phenomena that cause extensive damage to life and property. There are some safety measures that can be useful when one is faced with such disasters.

Safety During Volcanic Eruptions

If a volcano erupts near any settlement, it is advisable to evacuate that place and stay clear of lava, mudflows, flying rocks, and ash. The river areas and low-lying regions should be avoided. People should wear clothes that cover the whole body and cover their faces with masks. Doors and windows should be closed, and any vents in the houses should be blocked to prevent the entry of ash. People should also avoid driving cars and motorbikes, as ash can damage engines and metal parts.

▲ *A mask protects us from inhaling the smoke, ash, and toxic gases released during an eruption*

▼ *The caution board warns of the volcanic fumes ahead in a national park Kona Island, Hawaii*

Safety During Earthquakes

Earthquakes can cause death and destruction through secondary effects such as landslides, tsunamis, fires, and fault ruptures. The greatest losses in life and property occur due to the collapse of artificial and subsurface structures caused by the violent tremors of the ground. So, it is important to learn the safety measures and methods to prepare for an earthquake. People should be aware of the actions that they need to take when the tremors begin.

During an earthquake, the use of elevators and electrical gadgets should be avoided entirely. It is advisable to stand where nothing can fall on you. Everyone should keep a first-aid box at home. It is advisable to have dust masks, battery-operated radios, and flashlights in accessible places. Everyone should know to turn off the gas and other electrical appliances to avoid fires.

When the tremors begin, drop to the floor and take cover under a desk or table. Even if you are standing under a doorframe, hold on and do not leave its cover until the tremors stop entirely. Stay indoors until the tremors stop and you are sure that it is safe to exit. Even after this, stay away from bookcases or furniture that can fall suddenly. If you are outside when an earthquake hits, stay outside. Move away from buildings, trees, utility wires, and telephone poles.

▲ *In case of an earthquake, take shelter under a table or a desk*

What to do in case of an: EARTHQUAKE

What Is Weather?

Weather is the day-to-day condition of the atmosphere in a specific place at a specific time. It is nothing but a temporary state of the atmosphere that changes on an hourly or daily basis. It can be expressed in terms such as sunny, rainy, windy, or cloudy. Not all regions experience every type of weather. For example, a region in Africa under the desert climate will never experience snowy weather.

▲ *Weather is affected by the Sun along with the cloud and wind conditions*

What Causes Weather?

On Earth, weather is influenced by the heat of the Sun and the movement of air. There are certain elements that make up the weather of a place at a given point of time. They are temperature, precipitation, atmospheric pressure, wind, humidity, and cloudiness.

Temperature

Do you know what 'temperature' is? It can be defined as the measure of the hotness or coolness of any place. The angle of the Sun determines the increase or decrease in temperature. It can change quite quickly. Temperature influences the volume, pressure, and density of air. When temperature increases, air density decreases but its volume and pressure increase. So, warm and dry air will rise when it is surrounded by cool air because it is less dense than cool air.

Precipitation

Precipitation is the water formed in the atmosphere that falls on Earth's surface as rain, snow, and sleet. It is a result of the **condensation** of water vapour.

Atmospheric Pressure

The air that surrounds Earth has weight. This weight is termed as atmospheric pressure and it pushes downward on anything that is below it. The rise of warm air and fall of cold air cause changes in the atmospheric pressure. Rising air creates low pressure on Earth's surface, whereas sinking air creates high pressure. Therefore, atmospheric pressure is the highest close to waterbodies. For instance, islands and coastal areas are located close to waterbodies like oceans and seas. They frequently experience storms as a result of this, leading to constant and sudden changes in the weather.

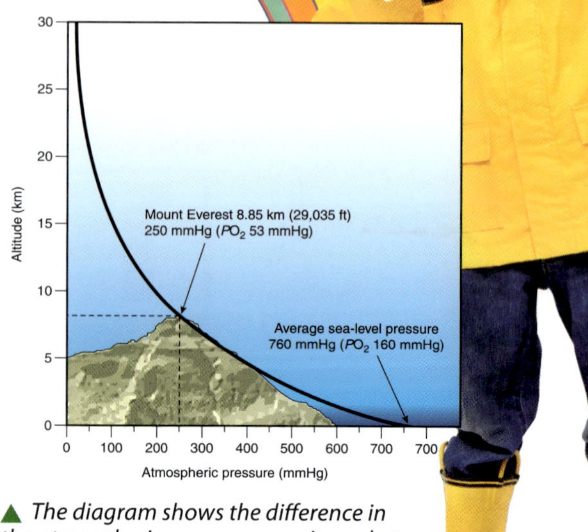

▲ *The diagram shows the difference in the atmospheric pressure experienced at sea level and high altitudes. Atmospheric pressure increases as we go closer to the water*

Cloudiness

Look up at the sky. Is it clear with few clouds, or is there a heavy cloud cover? What colour are the clouds? Cloudiness is the state of clouds in the atmosphere at a given time over a region. There are different types of clouds, and they affect the weather in different ways. For instance, fewer and whiter clouds indicate little or no rainfall. On the other hand, lots of dark, heavy clouds indicate heavy rainfall and even upcoming thunderstorms. Clouds can also stop the amount of sunlight reaching Earth's surface for a certain period of time. This makes the day cooler.

Isn't It Amazing!

Scientists believe that in one hour, 760 thunderstorms occur on Earth. Also, at any given second, Earth experiences an estimated 100 lightning strikes. Predictions about the number of thunderstorms and lightning have been made in the past, but they were inaccurate. With the use of satellites, scientists have been able to make more accurate estimates.

Humidity

Humidity is the amount of water vapour present in the air. It impacts the weather of a region. When the humidity is high, the day becomes hotter, and we begin to perspire. The level of humidity can help predict upcoming storms.

Wind

Wind is moving air formed due to differences in temperature and atmospheric pressure between nearby regions. Generally, winds tend to blow from areas of high pressure (where it is cooler) to areas of low pressure (where it is warmer). You might feel the wind against your face when it is moving. When the speed of the wind increases, the air moves faster, reducing warm air. This lowers the temperature and makes the weather cooler than normal.

▲ *The rain clouds appear to be grey because of their height and thickness. When the sky looks like this, it means that there might be some rainfall in the future*

Incredible Individuals

The anemometer was invented by Leon Battista Alberti in 1450. This device was used to measure the speed of the wind. It was a disk that was perpendicular to the direction of the wind.

Atmosphere: The Air Up There

There is an invisible layer surrounding Earth. It is composed of different gases. This layer is called the atmosphere. The word 'atmosphere' originates from the Greek words for vapour, '*atmos*' and sphere, '*sphaira*'.

A Gas Envelope

Earth's atmosphere contains oxygen and nitrogen in large quantities and then carbon dioxide, argon, helium, and neon in smaller quantities. Water vapour and dust are also a part of the atmosphere.

The atmosphere extends from the land, oceans, and snow-covered surface of the Earth right to the outer space. The gases in our planet's atmosphere are required for our survival and growth. Other planets also have atmospheres, but they contain different gases.

Incredible Individuals

Stratosphere accompanies all electrical storms. It was recognised as a chemical compound by German chemist, Christian Friedrich Schonbein in 1840. He named it 'ozone' due to the German word '*ozon*' which means 'to smell', as it had a pungent smell.

Layers of Atmosphere

The atmosphere is divided into layers just like the interior of Earth. There are five major layers—troposphere, stratosphere, mesosphere, thermosphere, and exosphere. Beyond the exosphere is outer space.

Exosphere
The exosphere is the highest layer of the atmosphere, extending 9,900 kilometres above Earth's surface from the thermosphere. It has the thinnest cover of air, gradually fading into space.

Thermosphere
The thermosphere is the fourth layer of the atmosphere. It extends for 690 kilometres from the mesopause, which is the boundary between the mesosphere and the thermosphere.

Space shuttles orbit in this layer. The temperature can rise to 1500° C here. The main feature of this layer is that it has free ions and electrons that are the result of the **ionisation** of gas particles. Radio waves bounce off the ionosphere, allowing communication between countries.

Mesosphere
The mesosphere is the third layer of the atmosphere. It lies 85 kilometres above Earth's surface. Air becomes thin in this layer as air molecules are further apart. There is a drastic fall in temperature here, as low as -120° C. Meteors burn out when they reach this layer.

Stratosphere
The stratosphere is just above the troposphere, extending from the tropopause to approximately 50 kilometres above Earth's surface.
This layer has small amounts of ozone, a form of oxygen that absorbs the harmful rays of the Sun, preventing them from reaching Earth's surface. The region in the stratosphere where this thin sheet of ozone is found is called the ozone layer. Aeroplanes and jet planes fly in this layer as it is calm and stable.

Troposphere
The troposphere is the first layer of Earth's atmosphere. This is the place where life can exist. It extends from the surface of Earth till the stratosphere. Its upper boundary is called the tropopause. It extends for 10–18 kilometres above Earth's surface. All the different weather changes occur in this layer. It contains almost all the water vapour of the atmosphere. Here, the temperature decreases as we go higher in altitude.

Ozone layer

The diagram shows the different layers of Earth's atmosphere, indicating the layer in which regular planes and jet planes can reach and also the one where we send our satellites!

What Is Climate?

Climate is the atmospheric condition of a particular place over a long period of time. It is the general pattern of weather that is recorded for 30 years or more. This does not mean that the area will only experience one type of weather. For example, a region that experiences a dry climate for most of the year can sometimes have rainy days, making the weather hot and humid for a short time.

Factors Affecting Climate

Different parts of the Earth experience different climates. Some places are extremely cold, some are very hot, and some experience rainfall throughout the year. The climate of any place is influenced by several factors. These are its **latitude**, **altitude**, distance from the sea, shape of the land or relief, ocean currents, and direction of winds.

Latitude

The climate of any region is determined by its latitude. How far is its distance from the Equator? This question will influence the temperature of any place, thus affecting its climate. The Sun's rays fall directly on the Equator and make the places close to it extremely hot. However, the Sun's rays have to pass through a thicker layer of atmosphere at the poles than the equator, making the climate near the poles cooler.

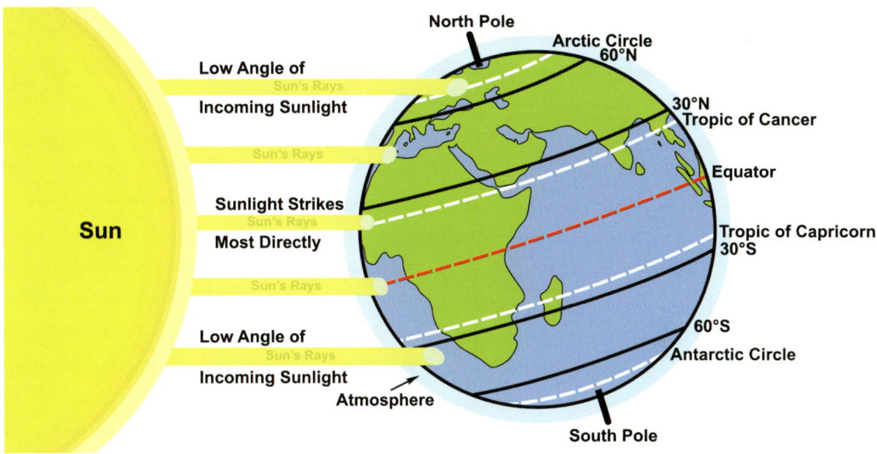

▲ *As the Sun's rays fall unevenly on Earth's surface, the climate varies*

Altitude

The climate of a place can be affected by its altitude. Mountains receive more rainfall than low-lying areas. As air is forced over the higher ground, it cools. This leads the moist air to condense and fall as precipitation. As we travel to places that have higher altitudes, we experience a fall in the temperature. Therefore, a place at sea level will have higher temperature than a place near tall mountains. Consequently, a place at sea level will experience a warm climate, and places on tall mountains at higher altitudes will experience a cooler climate. This fall in temperature has a cooling effect which causes condensation. It leads to precipitation on the windward side of the mountains, which is the side facing the wind.

▲ *Mountains affect the climate of a place*

▼ *Beaches are the preferred destination in summers because of their warmer climate*

Other Factors

Distance from the Sea

During the daytime, the Sun heats up the surface of the land and the ocean. Land warms quicker than the water in the oceans. This means that the air above land heats up quicker than the air above water. Warm air above land rises steadily, leading to low pressure near the surface. Conversely, there is high pressure over the water surface as the air is cooler. This cool air sinks lower and moves from high pressure to low pressure, cooling the land. This is called **sea breeze**.

The opposite of this process occurs at night. The warm air is now above the water. As the Sun goes down, the land loses heat quickly, cooling the air. However, the water in the ocean retains its warm air above it. Therefore, there is low pressure above the ocean and high pressure above the land. The air moves from land to ocean. This is called **land breeze**.

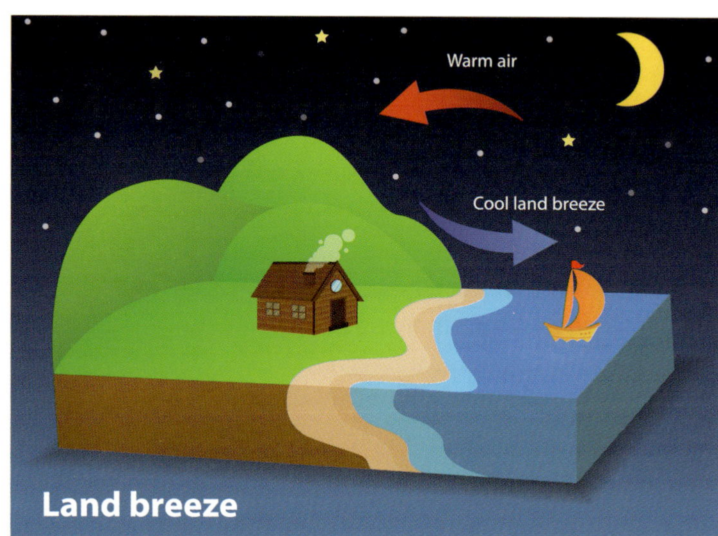

The sea breeze and the land breeze have a great impact on the climate of a place

Ocean Currents

If the wind blows in from the ocean, the warm and cold ocean currents can affect the climate of an area along the coast. Warm ocean currents heat the air above the water and carry warm air to the land, increasing the temperature of coastal areas.

Ocean currents take more energy to change temperatures than land. Land warms and cools faster than water

Direction of Winds

Winds blowing from the equator are warm. This makes the region hotter. Whereas, the cold winds from the polar regions make the adjacent areas cooler. Thus, winds can affect the climate of a place.

Types of Climates

Due to different factors, the world has varied types of climates. Some places are extremely cold, some have a moderate climate and some places are hot throughout the year. The climate across the globe has been divided into five climate groups—tropical, dry, mild, continental, and polar. These climate groups are further divided into climate types.

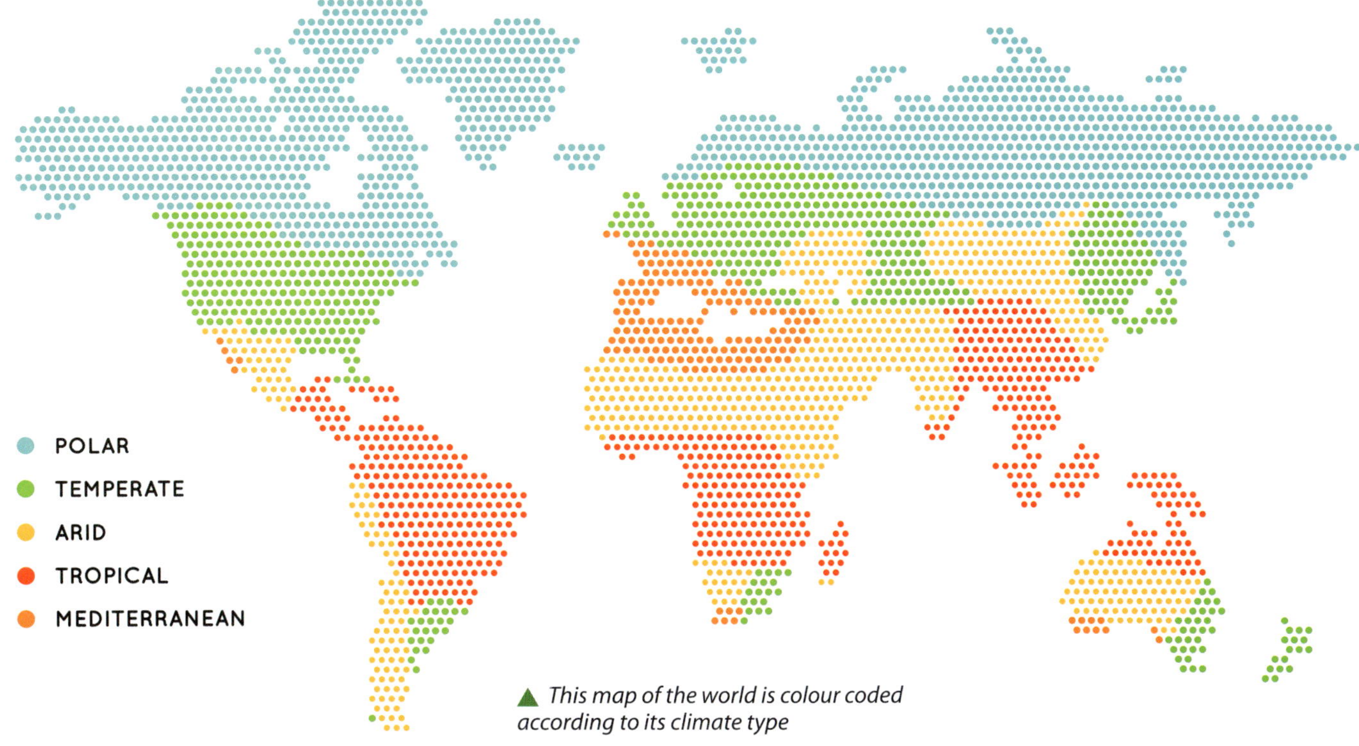

▲ *This map of the world is colour coded according to its climate type*

▲ *Tropical Brazilian rainforests*

▲ *The semi-arid region in north-eastern Brazil is also known as sertão*

How Climate Affects Us

Climate has a great influence on the development of any region. The crops farmers grow, the clothes people wear and the houses they build for themselves are all influenced by the climate of a place. Food habits depend on the type of crops that are grown in a particular region. People who live in cold climates wear clothes made from wool or fur. However, in places that are warm, people tend to wear light clothes made of cotton.

Incredible Individuals

How do we know the distribution of the climate types around the world? It is because of a Russian-born, German meteorologist named Wladimir Koppen. He prepared a map of the world demarcating the different climatic regions. He studied atmospheric science, contributing greatly to the subject of climatology and meteorology.

Climatic Zones

Earth is divided into three major climatic zones. They are the tropical or torrid zone, the temperate zone, and the frigid zone. The tropical zone lies between the Tropic of Cancer (23.5°N) and the Tropic of Capricorn (23.5°S). The temperate zone lies between the Tropic of Cancer and the Arctic Circle (66.5°N), and then again between the Tropic of Capricorn and the Antarctic Circle (66.5°S). The frigid zone lies between the Arctic Circle and the North Pole (90°N), and then again between the Antarctic Circle and the South Pole (90°S).

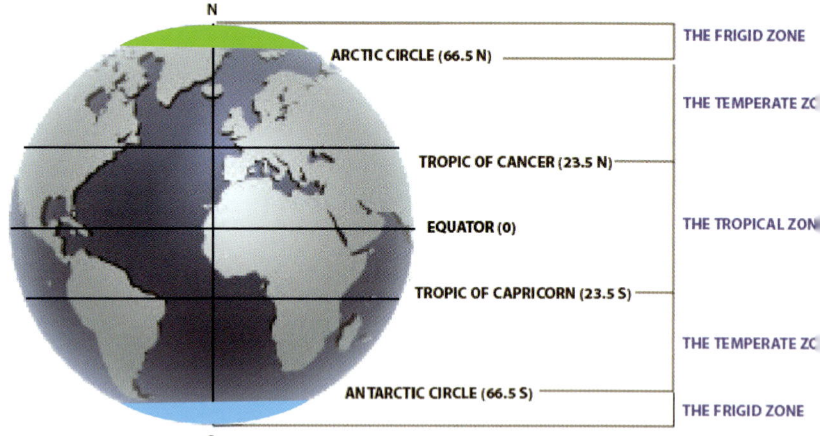

▲ *The major climatic zones are divided into the North and South Hemispheres by the Equator (0°)*

 The Tropical or Torrid Zone

The tropical zone has the Equator running through the centre. This region includes countries of South America, Africa, Asia, Australia, and parts of North America. The tropics have a warm climate round the year as these areas get the maximum heat from the Sun. The only two types of seasons experienced here are the dry and wet seasons. The amount of rainfall varies from place to place. The Sahara Desert in Africa experiences very little rainfall and is drier than the Amazon Basin of South America, where rainfall is plentiful.

▶ *The Congo Rainforest is a tropical rainforest*

 The Temperate Zone

The temperate zone has a broad range of temperatures and well-defined seasons. Weather is quite unpredictable in this zone. Some of the countries of the temperate zone are Canada, USA, India, and Japan. Countries of the Middle East, Northern Africa, and all the countries of Europe also fall under this climatic zone. The climate here varies widely. Some places have an oceanic climate as seen in London, Dublin, Melbourne, and Auckland. Some parts of the temperate zone have a Mediterranean climate. This is described as a dry summer climate.

◀ *The Kew Gardens in London is a botanical garden site hosting a diverse collection of plants that can grow in the temperate zone*

 The Frigid Zone

The frigid zone includes the polar regions. The climate at both the North Pole and the South Pole is extremely cold and freezing. The countries in the north polar regions are Alaska in the USA, Canada, Greenland, Norway, and Russia. Antarctica is in the southern polar region. This zone has cold temperatures as it lies far away from the Equator. Some areas are permanently covered with ice. The frigid zone receives slanting rays of the Sun as the region lies farthest from the equator.

Seasons

Seasons are the divisions in a year influenced by the changes in weather and climate. There are four seasons in a year—spring, summer, autumn or fall, and winter. Every season has its own characteristics. Seasons follow each other after a gap of three months. Each season is characterised by its own average temperatures, weather types, and durations of day and night. For instance, in summer the days are longer than the nights, while in winter the days are shorter than the nights.

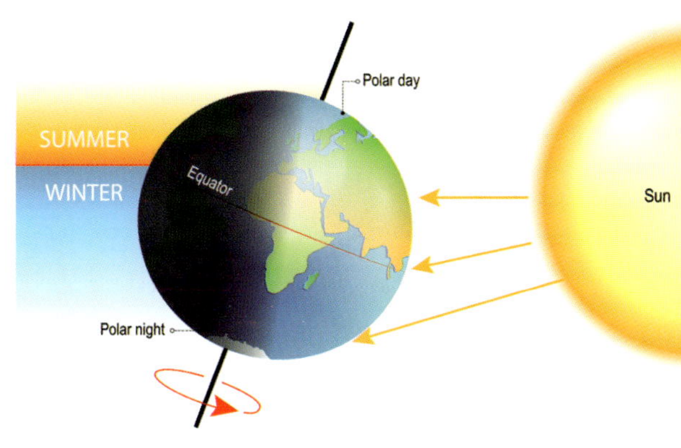

The Equator divides Earth into the Northern Hemisphere and the Southern Hemisphere

Spring

Spring is the season between winter and summer during which temperatures gradually rise and the climate becomes moderate. In the Northern Hemisphere it extends from the March equinox (20–21 March) to the summer solstice (20–22 June). In the Southern Hemisphere, spring begins from 22–23 September and lasts until 21–22 December. During spring, the changes in temperature occur mostly in the middle and high latitudes. However, near the Equator, temperatures do not vary much. The polar regions have very short spring seasons.

Spring is the time for most flowers to blossom

Summer is the most preferred season for outdoor activities

Summer

Summer is the season between spring and autumn. This is the warmest season of the year. It extends from the summer solstice to the autumnal equinox in the Northern Hemisphere and from December to March in the Southern Hemisphere. During summer, the days and nights are more or less equal in length.

Autumn or Fall

he autumn season comes in between summer and winter when temperatures begin to drop. This season is also called fall, as trees shed their leaves during this time. The leaves fall to the ground. Autumn extends itself from the autumnal equinox (22–23 September) in the Northern Hemisphere to the winter solstice on 21–22 December. This is the year's shortest day. Whereas in the Southern Hemisphere, it extends in the period between 20–21 March and 20–21 June. In the polar regions, the autumn season is very short.

Winter

Winter is the coldest season of the year. It falls between the seasons of autumn and spring. In the Northern Hemisphere, winter extends from the winter solstice on 21–22 December to the vernal equinox on 20–21 March. In the Southern Hemisphere, the winter season extends from 20–21 June to 22–23 September. The temperatures decrease considerably at this time in the middle and high latitudes of Earth. However, as we have seen for all other seasons, the temperatures in equatorial regions do not change much.

Water Cycle

Three-fourth of our planet's surface is covered with water. This water is constantly circulating in the air and on land. Water on Earth is available in all the three forms, i.e., as solid, liquid, and gas. The water cycle is a natural process by which water is transported from oceans into the atmosphere and then it returns to land. There are three main processes in the water cycle—evaporation, condensation, and precipitation. Clouds are made up of water droplets and crystals of ice. Snowflakes are made up of **ice crystals**. Rain is liquid water. Steam is water vapour. In this way, water exists in all three forms.

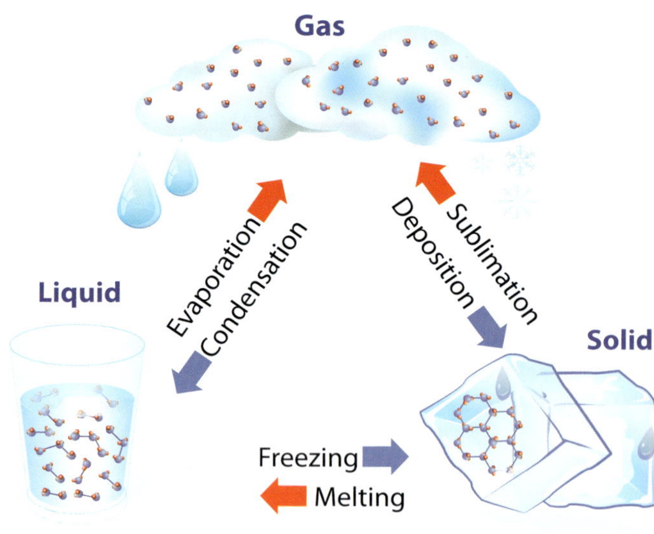

The transformation of water from one state to another

Evaporation

Evaporation is the process by which a liquid changes into gas. In the water cycle, water from oceans, seas, rivers, and lakes evaporates due to the Sun's heat. The water rises and forms water vapour. Water vapour is also formed directly from ice and snow in places that experience extremely cold weather with bright sunshine and strong winds. This is termed **sublimation**. **Transpiration** is another process by which water is transformed into vapour. This happens when plants release water onto the leaves, and then this water evaporates into the air as vapour. Some water reaches the land directly and flows across the ground. It gets collected in the waterbodies. This water is referred to as 'surface run-off'.

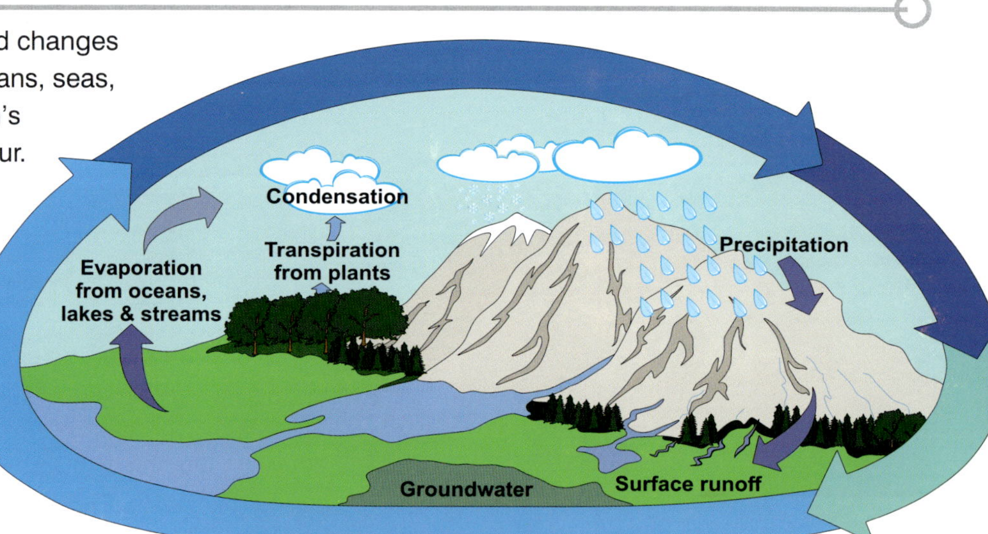

The continuous movement of water on Earth and its environment is called the water cycle

Condensation

Condensation is the process where a gas changes into liquid. It happens when water vapour in the air condenses and forms clouds either at higher levels in the atmosphere or even at ground level. This is a vital step in the water cycle.

Precipitation

Precipitation is any liquid or solid form of water that falls on Earth's surface due to condensation. When ample water gathers in a cloud, it becomes heavy. Droplets of water form and fall to the ground. Water that falls during precipitation either becomes a part of a body of water like an ocean or lake, or it seeps into the ground. Precipitation occurs in the form of rain, snow, sleet and hail (depending on the temperature of that place).

In Real Life

During the water cycle, water circulates in all its three forms—solid, liquid, and gas. The solid state of water is ice found in huge glaciers and ice capped mountains. This ice melts and flows as liquid, which is found in different water bodies on Earth. Due to the heat of the Sun, water from these water bodies evaporates into the air as water vapour, which is the gaseous state of water.

Predicting Weather

Weather forecasting is a field of meteorology in which a person tries to predict the weather. Weather forecasting is a scientific application by which one can predict the atmospheric conditions. The weather forecaster tries to describe what the weather of a place will look like for the next couple of days.

History

From the pre-historic age, human beings were able to detect changes in the weather by observing nature. They were able to spot the signs that predicted forthcoming snow, rain, or wind. This helped them move from place to place in search of food and shelter. In modern times, weather forecasts are made by meteorologists after the collection of information about the atmospheric conditions of a place through **meteorology**.

Meteorology and Meteorologists

Meteorology is the science of the atmosphere. It is derived from the Greek word for meteors meaning 'in the air', and '*logia*' meaning 'to discuss, study, and explain'. People who study meteorology are called meteorologists.

Meteorologists record temperature, humidity, wind speed, air pressure, and other weather patterns. They use this information to understand and predict the weather. They use mathematical models and computers to prepare their weather forecasts. Meteorologists also study the impact of the weather on the environment. They study climate and its patterns. Technology helps them understand long-term changes in the atmosphere.

Many organisations use weather forecasts for their benefit.

Thermometer

A thermometer gives a reading of the current air temperature. This reading gives information about whether the weather is cold or hot. A thermometer contains a thin tube of mercury that expands with heat and determines the temperature of a place. Today, digital thermometers measure the temperature electronically.

It must be noted that the thermometer used to measure body temperature is a different instrument altogether, though both the thermometers work on the same principle.

The three major temperature scales used in thermometers are Fahrenheit, Celsius, and Kelvin

Tools Used to Predict Weather

Meteorologists use a variety of tools to predict the weather and climate. Some well-known examples are thermometers, barometers, anemometers, computer models, and weather satellites.

In Real Life

During autumn, due to less light, there is no photosynthesis and leaves stop producing the green pigment called chlorophyll. As the green colour of the leaves begins to fade, orange and yellow **carotenoids** that are present in the leaf take the chance to shine. Therefore, the colour of leaves appears to be orange.

Tools Used to Predict Weather

Meteorologists use a variety of tools to predict the weather and climate. Some well-known examples are thermometers, barometers, anemometers, computer models, and weather satellites.

Barometer

A barometer is a device that is used to measure atmospheric pressure or barometric pressure. Like a thermometer, a traditional barometer indicates pressure on a linear scale or on a circular dial. However, these days, digital models also have an electronic display. A rising barometer generally indicates that the weather is clear and a falling one indicates that a storm is approaching.

▲ *The barometer was invented in 1643 by an Italian mathematician Evangelista Torricelli*

Anemometer

An anemometer is an instrument that measures wind speed. It is usually made of three or four cups attached to horizontal arms. These arms are attached to a vertical rod. As the wind blows, the cups rotate, making the rod spin. This rotation helps calculate the speed of the wind. Another type of anemometer is the windmill, which counts the revolutions made by windmill-style blades. The rod of windmill anemometers rotates horizontally. Because wind speeds are not consistent, they are usually averaged over a short period of time.

▲ *The first mechanical anemometer was invented in 1450 by Leon Battista Alberti, an Italian architect*

Computer Models

Meteorologists these days use sophisticated computer models to predict weather several days in advance. The model is a programme that predicts the intensity of air pressure and how high and low pressure zones will change over a given period of time. There are many models for meteorologists to calculate weather conditions.

Weather Satellites

There are two types of satellites that monitor the weather conditions on Earth—geostationary satellites and polar orbiting satellites. A geostationary satellite watches the weather over a particular region and provides continuous coverage of storms. This helps the meteorologist follow the same storm for a couple of days. The polar-orbiting satellite covers the whole Earth in about 12 hours. It takes images of many different regions to assist in better weather prediction.

▶ *Geostationary satellites orbit 35,786 kilometres above the equator, rotate at the same pace as Earth, and focus on the same area at all times*

Incredible Individuals

Daniel Gabriel Fahrenheit is famous for his invention of the mercury thermometer in 1714 that is widely used for measuring temperature. He also invented the alcohol thermometer in 1709. The Fahrenheit temperature scale, named after him, is still in use in the USA.

Changing Climate

Global warming today is a serious issue that needs to be addressed properly. Increased levels of greenhouse gases in the environment, mostly from human activities—such as the burning of fossil fuels, deforestation, and farming—cause climate change. While the Earth's climate is also influenced and changed through natural causes such as volcanic eruptions, ocean currents, the Earth's orbital changes, and solar variations, but human-led factors are speeding up several processes to the point of no return.

Natural Causes

Global warming is caused by the **greenhouse effect**. The atmosphere acts as a greenhouse and traps some of the heat of the Sun. Gases like carbon dioxide, water vapour, and methane behave as if there is a thin sheet of glass surrounding the planet. The Sun's rays pass through this layer of greenhouse gases and warm up the Earth. In turn, Earth lets off heat that radiates into space. Some of this radiation does not pass through the atmosphere but is reflected back to the Earth. The greenhouse gases in the atmosphere trap the heat and keep the planet hotter than it would otherwise be. This is called the natural greenhouse effect. Without this natural phenomenon, Earth would be much colder than it should be in order to support life.

Volcanic eruptions are a natural factor that affect the climate. The gases, lava, and dust particles emitted during an eruption can influence the climate. The particles spewed from volcanoes can cool the planet by shading it from incoming solar radiation. This cooling effect sometimes lasts for months or even years. It can change the climate of that particular region.

Human Activities

Industrialisation, burning of fossil fuels, and excessive use of vehicles has enhanced the greenhouse effect, which is a major cause of global warming. A large amount of carbon dioxide is emitted from factories, vehicles, and other human activities. Along with these, over the years, deforestation or the felling of trees has been another major cause of concern. Deforestation endangers wildlife and also increases the amounts of carbon dioxide in the atmosphere. Deforestation further contributes to greenhouse gas emissions through burning of forest biomass and decomposition of remaining plant material and soil carbon. The burning of fossil fuels, which at present provides 80 per cent of energy, is adding to global warming. This problem is only getting worse, and in the process Earth's temperature is also increasing, making life difficult on this planet.

Climate Change

Today, climate change is a different issue altogether. Scientists believe that the changes taking place are because of global warming. Changes have now become erratic and unpredictable, making some places hotter, some cooler, some drier, etc. In the polar regions, glaciers are melting due to climate change, and consequently sea levels are rising, leading to frequent floods and excessive erosion.

What should we do?

We should make every effort to stop the climate from changing so drastically. The best way to do so is to plant trees, try to conserve energy and water, and use electricity carefully. While each individual should take necessary steps to adopt sustainable habits, as a society we should seriously consider a shift to alternative sources of energy, and move towards more sustainable living—both personally and commercially.

ANIMALS

Animals, with their beauty, complexity, and ecological significance, form an essential part of the interconnected web of life on Earth, inspiring awe and wonder in all who encounter them.

They can be classified into various categories based on different characteristics, including birds, insects, invertebrates, mammals, marine animals, reptiles, and amphibians.

Birds are characterized by their feathers, beaks, and the ability to fly. They display a wide range of colors, sizes, and behaviors, from small hummingbirds to large ostriches. Insects, on the other hand, have six legs, a segmented body, and antennae. They are the most diverse group of animals on Earth, with millions of different species, including butterflies, ants, and beetles.

Invertebrates are animals without a backbone, such as spiders, worms, and jellyfish. They have a wide range of body structures and adaptations, allowing them to thrive in various environments. Mammals, on the other hand, are warm-blooded vertebrates that have fur or hair and give birth to live young. They include well-known animals like lions, dolphins, and monkeys.

Marine animals encompass a vast array of creatures, from tiny plankton to enormous whales. They have adapted to life in water and come in a variety of forms, such as fish, sharks, sea turtles, and octopuses. Reptiles are cold-blooded animals, characterized by scaly skin and laying eggs. They include snakes, lizards, crocodiles, and turtles. Amphibians, like frogs and salamanders, are vertebrates that spend part of their lives in water and part on land.

The diversity of animals is truly remarkable, with each group exhibiting unique features and adaptations that enable them to thrive in different habitats and fulfill various ecological roles. Through their distinctive characteristics, animals contribute to the richness and balance of our natural world.

Frogs are amphibians as they spend parts of their lives on land and in water

Although an alligator is a reptile, it can swim well in water and crawl on land

Birds: Then and Now

You might have seen pictures of dinosaurs that roamed over Earth long before our time. Did you know that scientists have found **fossils** of dinosaurs with bird-like features? This has led them to believe that dinosaurs were the ancestors of birds. It is said the ancestors of birds were theropod dinosaurs. The famous *Tyrannosaurus rex* belongs to the same family.

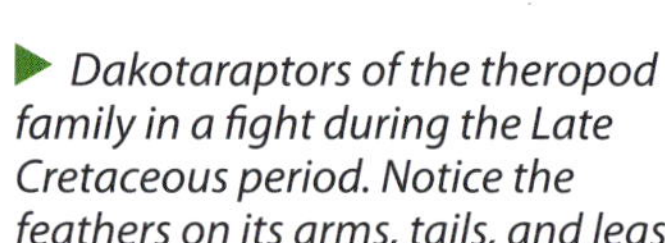

▶ *Dakotaraptors of the theropod family in a fight during the Late Cretaceous period. Notice the feathers on its arms, tails, and legs*

▲ *Modern birds evolved in the Jurassic and Cretaceous eras alongside dinosaurs*

Incredible Individuals

Sir David Attenborough is a British naturalist and broadcaster. He filmed various animals in their natural environments and educated viewers on how they behave in the wild. He has received several awards for his programmes.

The Missing Link

Have you heard about the Solnhofen Limestone? This is a geographical formation found in Germany. Its speciality is that many fossilised organisms were found here.

In 1861, a startling discovery was made here, which later came to be known as the missing link between dinosaurs and birds. A fossil of a creature named *Archaeopteryx* was found. It was first classified as a bird, but after finding fossils of similar bird-like creatures, it was reclassified as a dinosaur in the 20th century.

The *Archaeopteryx* had small teeth similar to a dinosaur and wings and feathers, similar to a bird. It lived about 147 million years ago in the Late Jurassic period. It was the size of a big hen, but with a light body that could glide through trees.

The Modern Birds

As time passed, toothed mouths were replaced with toothless and horn-shaped beaks. The *Confuciusornis* whose fossils were found in China, was the first bird to have this type of beak. It lived more than 100 million years ago and was about 25 centimetres in length.

Another bird-like creature called *Ichthyornis*, which lived 82–87 million years ago, was discovered to have a light skull with a beak. It resembled the modern seagull in how it had strong wings, (possibly) webbed feet, and ate fish. Nearly 66 million years ago, an asteroid hit Earth and wiped out the dinosaur species. However, the early birds, which were smaller in size, survived. This was because of their size and ability to adapt to the environmental changes of those times.

The first modern-day birds called the *Neornithes* emerged nearly 60 to 90 million years ago. In the next few million years, the number and types of birds increased. A journey that is believed to have started over 150 million years ago, culminated with trials and errors to create the beautiful birds we see today.

▲ *Fossil of Confuciusornis with a clearly marked beak*

Isn't It Amazing!

About 60 million years ago, South America was an island. This is where the terror birds roamed. One species of terror bird, called *Brontornis*, weighed up to 400 kilograms and ambushed its prey. Another, *Titanis*, could grow up to three-metres tall. With the continental drift, the North and South American landmasses joined at the Isthmus of Panama 2.8 million years ago. Climate change and competition from southward-migrating predatory mammals like wolves and sabre-toothed tigers drove these birds extinct in a few hundred thousand years.

Successful Evolution

When dinosaurs were wiped out of the planet, the mammals were still small and could not compete with the massive 'terror birds' that roamed the land. The birds around this time were so large that some even had a wingspan of 20 ft. The unnamed bird shown here is said to be the biggest bird that ever existed. They might have been the 'elephant birds' who are linked to modern ostriches.

Why were birds so successful? It is because they had the ability to fly. They could take off, successfully defending themselves from enemies. They were also able to reach farther than mammals to find food and shelter. Birds have met the trials of evolution so perfectly that Sir David Attenborough, a popular broadcaster and historian believes them to be the most successful creatures on Earth!

▲ *These scary birds disappeared around 2.5 million years ago*

◀ *Terror birds reached nightmarish proportions, and usually ranged from one-three metres in height and 350–400 kilograms of weight*

A Peek at the Birds

Have you ever observed birds closely? Can you think of all the different parts that give them the incredible ability to fly? Let's find out about all the important parts of a bird's body.

▼ Macaws are the largest type of parrots

Warm-blooded

Birds, like mammals, are warm-blooded in nature. Their bodies maintain a steady temperature regardless of the temperature of their surroundings. They do not rely on sunshine to increase heat within them during winter, or on shade to decrease the heat during summer. Usually, their bodies maintain a temperature of 40° C.

◀ The blue and yellow macaw is found in the rain forest in South America

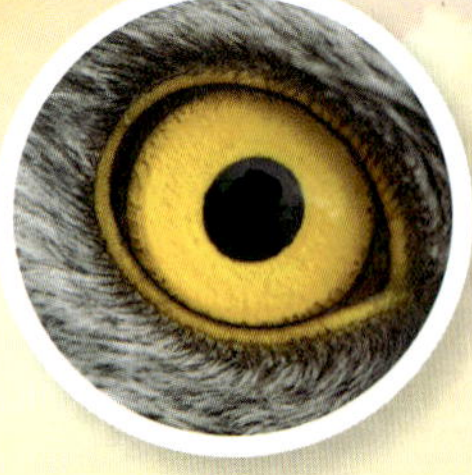

▲ An owl's eyes are adapted to seeing in the night

Aerodynamics

Birds' bodies are adapted to fly. Their bones are light and hollow. Their legs are held close to their bodies, giving them an **aerodynamic** advantage. Birds have no external ears. This feature streamlines their bodies, allowing them to fly with very little resistance. Also, importantly, birds do not sweat as they lack sweat glands. This keeps their feathers light and dry.

In Real Life

The next time you visit a zoo, pay close attention to the owls. Observe their eyes when they blink because they appear to have a third eyelid. This is nothing but a transparent membrane that moves across the eye in most birds. It is called the nictitating membrane. It protects the eyes from injury.

Bird's-eye View 

Birds have keen vision, especially birds of prey such as hawks, eagles, vultures, etc. Their eyes help them spot food from great distances. Birds of prey such as the eagle can see as far as two kilometres. Their eyes face forward, which helps them determine the distance between themselves and the prey. Birds that are preyed upon such as sparrows or pigeons—have wider and better fields of vision, as their eyes are on the side.

Beaks are special to birds and they come in various sizes and shapes depending on the food they consume

Beaks and Bills 

Birds have no teeth or jaws, but they have beaks, also known as bills. Beaks are primarily used to catch and eat food. They are also used to drink water, feed the young, gather materials to build nests, and for both offence and defence.

Bird Sounds 

Can you imitate the sound that a crow or a peacock makes? Birds make different, often melodious, sounds. They make calls, which are short tunes. Or they may sing in long, repeated patterns.

Birds call out when they sense danger from a predator or when they are angry, worried, or guarding their territory. They also sing when they are happy or looking for a mate.

The orange-bellied flowerpecker has very light feet which help it to hold onto branches of trees

Adaptations

Birds have feet which are adapted to where and how they live. A hen has feet suited to scratching the ground to find insects and seeds to eat. Commonly seen birds like pigeons, sparrows, and crows have thin feet with three toes facing frontwards and one backwards.

These types of feet help the bird perch and hold onto tree branches tightly. On the other hand, birds of prey like the eagle or vulture have strong feet with curved talons to kill the prey.

Incredible Individuals

The study of birds is enhanced by artists like John James Audubon (1785–1851) who, in his lifetime, painted a large number of birds existing in North America. He published a book titled *Birds of America* which had seven volumes.

Of Beaks and Birds

Birds use their beaks in many ways. They collect grains to feed their young and gather materials to build a nest, attack and defend, attract their mates, scratch themselves, and also use the beaks for grooming. Beaks can be classified according to size and shape as they help birds follow a specific diet.

Toucans have long bills that help them attract mates

Beak Skin

A bird's beak is covered with skin. The skin produces **keratin**, a fibrous, structural protein found even in the feathers of a bird. It is the same protein found in horns, hooves, hair, and fingernails across different species- including humans. The keratin dries up, making the beak hard and tough. It is this protein that gives a shiny appearance to the beak. Few birds like ducks, geese, and swans have a hard, flattened, horny tip at the end of their beak. It is called the nail, and these birds use their nail to dig into mud or swamps for food.

Parts of a Beak

The upper part of the beak is called the upper mandible. This grows out of the bird's skull. It cannot move independently. The lower part of the beak is called the lower mandible. It can move independently just like our lower jaw. Beaks come in various shapes and sizes and each one is suited to fulfil the needs of that bird.

Water-sifting Beaks

Most ducks, like the mallard, have a flat, broad beak which they dip into the water. When ducks are thirsty or hungry, they take in mouthfuls of water, which contains food in the form of insects, aquatic plants, algae, and small fish. Their beaks have tiny projections which look like the teeth of a comb. These projections help the duck sort out what they want to eat from the water.

Meat-eating Beaks

Birds of prey like the eagles, vultures, owls, and falcons have hooked beaks. These beaks not only help them to swoop down on the prey, but also to pull off the skin, fur, or feathers of the prey; and to tear apart the meat into small bites which they can swallow.

Ducks have big bills, which is a name for more fleshy beaks

Nectar-feeding Beaks

Hummingbirds have long, thin, needle-like beaks which they use to suck nectar from flowers. The beak is a protective covering for the tongue inside the mouth. It is this tongue that is used by the hummingbird to pull out nectar from the flowers.

A pouch-like beak helps capture food better

Different hummingbirds have beaks of varied size. The sword-billed hummingbird's beak can go up to 10.2 centimetres, which is longer than its body if you exclude the tail!

Fish-eating Beaks

The pelican has a huge beak with a throat pouch. The bird takes in the fish along with water into the pouch. The water is drained out and the fish is swallowed. The upper mandible in the pelican's beak has a small hook-like structure. This is used to spear the fish.

Parrots are often called 'hookbills', based on the shape of their beak/bill

Fruit-and-nut-eating Beaks

Birds that eat fruits and nuts, such as parrots or macaws, have smaller beaks with hooked tips. These tips are used to rip off the skin of fruits and reach the fleshy and sweet interior. These birds also use their beaks to break tough nuts down to edible pieces.

Seed-eating Birds

Seed-eating birds use their short beaks to pick up seeds to eat. Sparrows, like most such birds, have cone-shaped beaks that they use to peck on seeds.

The Gouldian finch is a seed eater

A beak with a hook gives a good grip to catch prey

In Real Life

Birds use their beaks to feed their young. An eagle captures prey and breaks it into small chunks. The eaglet (as its baby is called) snatches the feed from its parent and swallows it. Most eaglets eat one to eight times a day.

The baby bird that screeches louder than others usually gets more food

Colourful Feathers

A unique feature of birds is their ability to fly. It is enabled by feathers and wings. Most birds have bright and colourful feathers with striking patterns. Besides flying, they use their feathers to keep warm in winters, keep cool in summers, camouflage themselves from predators, and attract mates.

Inside Feathers

Feathers are made of a fibrous structural protein called keratin. Each feather consists of a hollow shaft and two vanes or two halves on either side. The vanes are made up of hundreds of thin branch-like barbs. On both sides of each barb, there could be still smaller branches called barbules.

A Variety of Feathers

Feathers are roughly divided into six types—contour, flight, down, filoplume, semiplume, and bristle.

Shed Me!

Birds shed feathers and replace them with new ones. This process is called moulting. It helps keep the feathers in good condition as it replaces the damaged and worn-out ones. Moulting is a time- and energy-consuming process. Hence, birds usually moult when they are not nesting or migrating.

How regularly do birds moult? Do they shed all their feathers at once? Some birds like eagles moult gradually and over a period of time so that they do not lose all their feathers at a time, and are capable of flying. On the other hand, birds like ducks lose their flight feathers immediately after the nesting season. During the few weeks it takes for the feathers to grow back, they remain flightless. But they do not go hungry since they find food by walking or swimming.

Isn't It Amazing!

Can you imagine rubbing ants all over your body? It would hurt your skin! But birds like crows like to go 'anting' during moulting. It means to pick up ants with their beaks and rub them on their feathers and skin. The birds may also lie near anthills to allow the insects to crawl all over them. They do this because secretions from the ants soothe their irritated skin during moulting.

▲ *This bird has contour feathers on its tail and its wings. These feathers give it a streamlined shape*

▲ *Many insect-eating birds need bristle feathers for feeding*

▲ *Down feathers have a fluffy appearance as their barbs are not joined together*

Play of Pigments

Feathers get their colours either due to pigments or because of light refraction caused due to the shape of the feather. Pigments are naturally colouring substances in plants, birds, and animals. In birds, pigments come in three varieties called melanin, carotenoids, and porphyrins.

Melanin

Melanin is found in both the skin and feathers of a bird. It produces colours such as black, brown, and yellow. Melanin is also known to make feathers strong and resistant. It is the primary pigment in birds such as crows, owls, and hawks.

Porphyrins

Porphyrins are pigments that can produce pink, green, and brown colours in birds. Some species of hummingbirds and peacocks get their shimmering colours from this pigment and the structure of their feathers. The colours are produced because of light refracted by the protein in the feather. The colours might vary depending on the angle at which the feather is viewed.

▲ *A white-tailed ptarmigan is also referred to as snow quail*

Carotenoids

Birds cannot produce carotenoids but they acquire them from the food they eat. These pigments produce colours such as bright red, orange, and yellow. The flamingo looks pink, but it will lose its colour if it does not eat carotenoid-rich foods like shrimp and algae. Goldfinches and cardinals also need carotenoids.

▲ *The Guinea turaco has a green plumage because of porphyrins*

Hide Me, See Me

Camouflage is used to conceal or blend oneself with the surroundings for safety. Birds camouflage to protect themselves as well as their eggs and later the young ones from predators. Colourful feathers are the best tools for disguise; the white-tailed ptarmigan is a great example of the same. In winter, the feathers of this bird—that resides in high altitudes—turn snow-white to blend in with the snow. In summer, the bird is streaked grey and brown. However, its tail remains white throughout the year.

In Real Life

Have you seen a bird preening? It refers to the process of cleaning and tidying the feathers. A bird preens to make sure the feathers work properly. For this, it squeezes out oil from the oil gland under its tail. With the help of the beak and claws it rubs this oil along the length of the feathers. The feathers become shiny, smooth, and waterproof.

▼ *Peacocks take three years to grow their tail feathers*

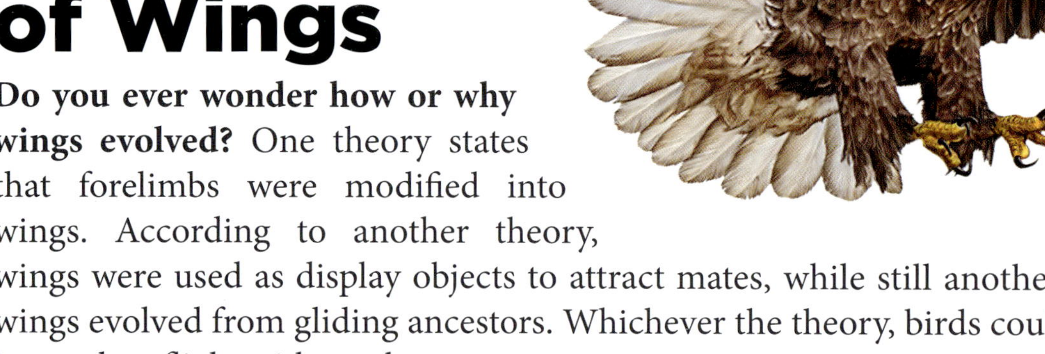

The Wonder of Wings

▲ The huge wings of eagles help them soar and glide with minimal effort

Do you ever wonder how or why wings evolved? One theory states that forelimbs were modified into wings. According to another theory, wings were used as display objects to attract mates, while still another says wings evolved from gliding ancestors. Whichever the theory, birds could not have taken flight without these structures.

Is it a Bird? Is it a Plane?

The wings of an aeroplane are modelled on bird wings. Wings are attached to powerful chest muscles. They are convex on the upper surface, concave on the lower, and they taper from front to back.

When a bird flies, air flows at greater speed at the upper part of the wing than the lower part because of its shape, creating less pressure on the top. This helps lift the bird during flight.

How Birds Use Wings

There are different types of wings based on the flight of the bird. Birds of prey like eagles, vultures, and hawks have soaring wings that are large and broad. The moment the bird spots prey, it closes its wings and dives down, opening them again to slow down as it nears the ground to make a soft landing.

◀ A Cape vulture has a wingspan of 2.26–2.6 metres

▶ Adult European gulls have white heads and bodies and grey wings

Birds such as seagulls and Arctic terns have gliding wings. These are long, narrow, and flat, with no space between the feathers. Gliding wings allow the birds to fly without having to spend much energy to stay in the air. Gliding refers to a bird moving in a downward direction closer to land and thus it must occasionally flap its wings to regain height.

◀ The flying sparrow has high-speed wings

Small birds such as house sparrows, woodpeckers, and thrushes must be quick to escape their predators. They have elliptical wings. These rapid take-off wings have spaces between the feathers, making them lighter and easier to move. These wings are not made for sustained flight and the birds must expend a lot of energy to stay airborne for long periods of time. Scavenger birds such as crows and ravens also have elliptical wings that help them steal food and escape quickly.

Birds such as swallows and swifts, which must catch their food mid-air at times, have high-speed wings. These wings are long, narrow, and pointed. They are angled backwards, making rapid flight easy.

The No-Fly List

It is not true that all birds fly. Even though they have wings, birds like ostriches, emus, kiwis, rheas, and cassowaries cannot fly. Their flat breastbone, to which the strong flight muscles are attached, lack a keel. This means that their wings are weak and cannot take off the ground. Flightless birds as a group are called **ratites**.

Long Lost Cousins

It is believed that ratites are related to a group of 47 South American birds called 'tinamou'. While ratites lost their ability to fly over time, tinamous are capable of flight, but prefer to walk on the ground.

Scientists believe that over millions of years, ratites lost flight as a trait as they adapted to their environment. Wings became a redundant feature. However, the birds evolved strong muscular legs instead, making them fast runners and offering an alternative way to escape predators.

World's Biggest Bird

The ostrich is the largest living flightless bird. It is native to the continent of Africa. In fact, even the egg the ostrich lays is the largest of all eggs, each about 6 inches in length, 15-18 inches in circumference, and around 1.4 kilograms. When it senses danger, the bird shoots off at a speed of 72 kmph and while facing a threat, it is known to kick hard.

The Australian Wonder

The emu is the second-largest living bird in the world. It is about 5 feet tall and is found on the continent of Australia. Like the ostrich it runs very fast, at 48 kmph, and kicks at the predator.

South American Rhea

Rheas are found in South America. They are omnivores, which means that they eat both plants and animals. They are related to the ostrich and emu—who live in different continents. But, they are much smaller, measuring 4 feet in height and 20 kilograms in weight.

Rheas are of two types—the common rhea which is found in Brazil and Argentina, and its cousin, which is a slightly smaller bird called Darwin's rhea, which lives in areas starting from Peru, all the way down to Patagonia.

Isn't It Amazing!

When dinosaurs were wiped off the planet, mammals were still small and had not yet evolved to become apex predators. Instead, birds underwent what is known as an 'evolutionary explosion'—occupying several ecological niches. Some, like the hummingbird, evolved to be very tiny and live on nectar, while others like the giant moa became large herbivores. Birds were successful because they could fly and take off at will, successfully defending themselves from enemies. They were also able to reach farther than mammals to find food and shelter.

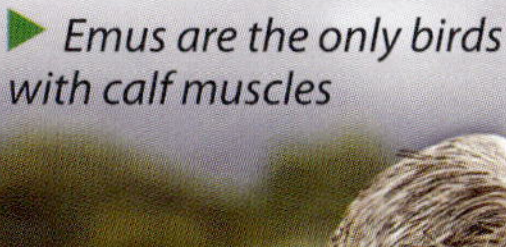

Emus are the only birds with calf muscles

The adult ostrich is 9 ft in height

Ostriches have two toes, but rheas have three

One ostrich egg is equivalent to 24 hen's eggs

Spot the Insect

Insects account for almost 80 per cent of the animal species in the world. Now that is a huge number! All of these insects belong to the larger group of invertebrate animals known as **arthropods**. This is the same group to which scorpions, spiders, crabs, and lobsters belong.

Body

An adult insect found anywhere in the world has three common body parts—head, thorax, and abdomen. Most insects have three pairs of legs, two pairs of wings, and two antennae. The eyes, mouth, and antennae are on the head; the legs and wings are attached to the thorax. The abdomen has most of the organs, such as the stomach and intestine.

The jewel beetle is one of the most colourful insects in the world, belonging to order Coleoptera

Exoskeleton

Insects are covered with a hard exoskeleton made up of chitin, which contains proteins and carbohydrates. In an insect, the eyes are covered with a thin layer of exoskeleton. This is a non-living structure; hence it does not grow with the insect. At every stage of growth, the insect moults.

Moulting refers to the process by which the exoskeleton is shed to regrow a new one. At this point of time the insect, with its soft exterior, is most vulnerable to predators and its surrounding environment. Many insects go into hiding until the tough, new exoskeleton grows back.

Apart from protection, the exoskeleton of the insect gives its body shape and prevents it from drying. It has tiny holes called **spiracles**. These are small openings or pores that allow respiration. The exoskeleton is made up of varied patterns and colours. Bright colours warn the predators that the insect is poisonous, while dull colours make for good camouflage.

A cicada shedding its exoskeleton while entering the next stage of its life

Classification

All insects belong to the class Insecta. They are classified into several groups. Most of the insects on our planet belong to the following orders—Coleoptera (scarabs and beetles), Lepidoptera (butterflies and moths), Hymenoptera (ants, wasps, and bees), and Diptera (flies or 'true' flies).

There are many species of insects belonging to each of these classes. 40 per cent of the known species of insects belong to order Coleoptera. The order Lepidoptera has about 180,000 species of insects. Order Hymenoptera has about 115,000 known species, while around 125,000 species belong to order Diptera.

Sense-able

Insects have a good sense of touch, smell, hearing, and sight. They do not have ears, but their skin can pick up sound waves. For a few insects such as grasshoppers and crickets, the tympanal organ acts like the ears. Insects use their antennae to feel and smell. They need to have a keen sense of smell to detect food and mates.

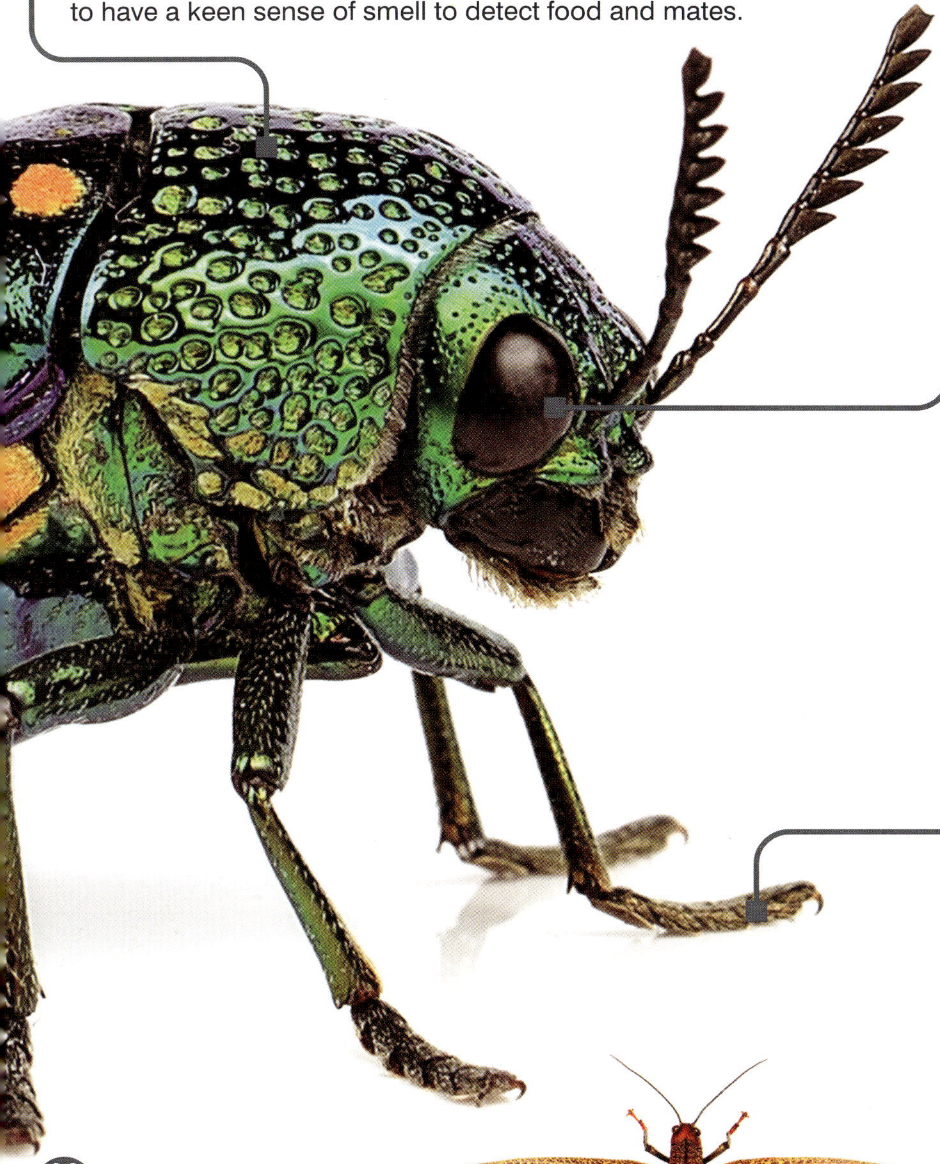

Eyesight

Insect eyes can detect movement but cannot see shapes as we do. Not all insects can see colours, but few, such as butterflies have exceptional vision to find flowers.

Insects can have two types of eyes; each species might have either of the types or both. The first type are called the ocelli. These are simple, small eyes. They cannot see well but can detect dark and light. These eyes are seen on insects like fleas.

The second type are huge, bulging eyes like that of the housefly. These are called compound eyes. They are made up of many tiny single eyes. Each single eye is called the ommatidium. It sees its own image and sends it to the insect brain. The brain then combines all the images to form a bigger picture.

▲ *Wasps and ants have both types of eyes*

Of Legs and Joints

The word 'arthropod' means to have jointed feet. Insects are arthropods, which means their legs have joints just like the joints on human legs. However, their legs do not have bones inside them; they have muscles instead. These muscles allow the insects to move, bend, run, and jump.

One of the fastest insects in the world is the Australian tiger beetle. It cannot fly but can run at a speed of 9 kmph. Another fast insect is the tiger moth caterpillar which picks up a speed of 5 kmph.

Wings

Insect wings are made up of chitin. The muscles attached to the exoskeleton make these wings flap. Most insects such as bees, beetles, and dragonflies have two pairs. Insect wings are thin and transparent, but veins running between them make them sturdy.

▲ *A species of grasshopper called tropidacris cristata*

This is not to say that all insects have wings. Remember, there is great diversity in the insect world. Fleas, lice, and bedbugs, for example, are wingless. Few insects such as aphids can even grow wings whenever needed, such as in the case of food shortage or overcrowding. At such times, the insects need wings to fly away to a better place. Aphids are commonly called greenflies and blackflies. Many aphid species choose to feed on only one type of plant for their entire lives. So, they are called **monophagous**.

Isn't It Amazing!

Cicadas are among the loudest insects in the world. They have special organs called tymbals that produce their characteristic sounds which can sometimes reach 90 decibels! This is as loud as the sound produced by a motorcycle.

The Menacing Weevil

Beetles and weevils belong to the order Coleoptera. This order has close to 350,000 insect species. Coleoptera is not just the largest insect order, but also the largest group of animals in the world. The number of insects classified within this order is rising each year with the discovery of new species by scientists. Coleopterans are seen in almost all habitats of the world, right from oceans, mountains, and deserts to cold regions.

Colourful Beetles

The word 'Coleoptera' comes from the Greek words, '*koleos*' or sheath and '*pteron*' or wings. This refers to the special arrangements of wings in beetles. Similar to other insects, beetles have two pairs of wings. However, one pair is slightly modified.

The front wings in these beetles are hard and strong. They cover the upper side of the body like a sheath. This hard casing-like structure is called the **elytra**. Elytra protect the delicate rear wings. The rear wings are folded under the elytra at rest, but when the beetle is about to fly, they emerge from the casing.

▲ *Beetles exist in different colours. This is the spotted cucumber beetle*

Weevils

There are about 40,000 species of weevils in existence in the world. Most species are brown or grey, but the diamond beetle is colourful. Weevils are also called snout beetles because they have long, curved snouts which look like elephant trunks. The snouts are used to make holes in leaves to lay eggs. They are also used to penetrate the leaves.

Most weevils have long foldable antennae. These antennae fold into special grooves present on the snout. They are small animals, measuring about 6 millimetres; but there are exceptions with few measuring almost up to 7–8 centimetres. Weevils feed on fruits, stems, flowers, and seeds.

▲ *Weevils have managed to survive because of their snouts*

Into the Grain

Grain weevils are small brown weevils, not more than 4 millimetres long. They make the most menacing pests as they destroy stored grain such as maize, oats, and wheat. The females bore holes in individual grains to lay eggs. The larvae, which emerge from these eggs, feed on the grain. When they are threatened, they feign death. This is how they protect themselves from predators.

▲ *Grain weevils are also known as wheat weevils*

Cotton Soft

The boll weevil is about 6 millimetres in length. But it makes for a terrible guest, as it enjoys destroying its host—the cotton bolls. The female boll weevil lays eggs in cotton buds. The larvae from the eggs feed on the cotton seeds as well as fibres, completely destroying them. Farmers have to bear immense losses every year because of the boll weevils.

▲ *Boll weevils cannot feed or breed on any plant other than cotton*

Butterfly or Moth?

Butterflies and moths belong to order Lepidoptera, whose name means 'scaly wings'. The large, thin wings of these insects are covered with tiny scales. Along with the wings, the body and legs too are covered with scales, which come off if the insects are held. The order Lepidoptera is the biggest family of insects after Coleoptera. It consists of almost 180,000 species.

Similar but Different

There is a reason that butterflies and moths look similar—they are cousins! However, they have many differences which make it easy to tell them apart. There are 20,000 species of butterflies, but there are a whopping 160,000 species of moths. While butterflies are active during the day, moths are active during the night. Butterflies sport bright and beautiful colours. But moths often have dull brown or grey patterns on their wings. Butterflies have club-like antennae, while moths have feathery antennae.

Sweet Nectar

What do butterflies and moths feed on? They eat the nectar produced in flowers. Nectar is a food source containing lots of energy-rich sugars. Butterflies and moths eat nectar when they become adults as they need lots of energy to fly.

▶ *Butterflies come in various colours and patterns. Several of them hide from predators using their colourful wings*

Usefulness

Lepidopterans have made homes in all continents except Antarctica. They live in forests, grasslands, mountains, deserts, and even cities.

Butterflies and moths are important **cross pollinators**. For example, the South American cactus moth has been introduced in Australia to clear out hectares and hectares of prickly pear cactus, a weed harmful to crops.

Trivia

The peacock butterfly has a special pattern on its wings that looks like eyes. It uses these eyespots to scare and ward off predators.

The sunset moth found in Madagascar is often mistaken for a butterfly. This is because it is active during the day and has colourful patterns on its wings. It is considered to be one of the world's most beautiful insects.

In Real Life

Moths are nocturnal insects. They are attracted to light because of a phenomenon called phototaxis. The movement of their wings is influenced by the strength of light. With distant sources of light, such as the Moon, the light reaches equally to both eyes, thereby causing the insect to fly in a straight line. But if the source of light is closer, such as a candle flame or electric bulb, the moth perceives it strongly in one eye rather than in both eyes. As a result, the wings on one side are stimulated to move faster, causing the insect to fly right into the light source.

▲ *Many adult moths don't eat*

Jump and Chirp

Crickets and grasshoppers belong to the insect order Orthoptera. This is the same order that dragonflies, stoneflies, and cockroaches belong to. Most of these insects have the ability to camouflage so that they can blend into their surroundings.

▲ *Most grasshoppers have a herbivorous die*

The Hind Wonder

Have you ever tried to touch a grasshopper? If you have, you must have noticed the leaps it takes. Both crickets and grasshoppers have powerful hindlegs for leaping and jumping, usually to escape predators. As they leap, they spread their rear wings. The rear wings are larger, with a membrane, and are usually colourful. Both the insects have rigid forewings.

Sing a Song

The chirping of crickets and grasshoppers is a very familiar sound in summers, especially in cold regions. The noise created is not actual singing but is called **stridulating**. It is created by rubbing two body parts together. Crickets rub their front wings to create the sound, while grasshoppers rub their hindlegs against the front wings.

◀ *Crickets have more protein than beef or salmon*

The Ear Tale

Grasshoppers and crickets have no ears. Instead they both have an organ—a stretched membrane of skin—called the tympanum. In case of crickets, it is located on the knee-like joints of their front legs, while in grasshoppers the membrane is located underneath the abdomen. The organ detects sound vibrations which are sent to the brain.

Crickets v/s Grasshoppers

Crickets and grasshoppers have certain dissimilarities which make it easy to tell them apart. Crickets have long antennae, while grasshoppers have short antennae. Crickets come out at dusk, but grasshoppers are active during the day. Crickets do not eat plants, but they can eat animals and each other. On the other hand, grasshoppers strictly rely on plants for food. An earwig is a small herbivorous, nocturnal insect belonging to the same family as the grasshoppers and crickets. It sprays a foul-smelling liquid to defend itself from predators, but it is harmless to us.

The Praying Mantis

The praying mantis, also known as *Mantis religiosa*, is a small but fierce hunter. It is known to take on animals much larger than its own size.

What's in a Name?

The praying mantis gets its name because of its long front legs. They are bent such that the mantis seems to be kneeling in prayer. This insect belongs to a bigger group of hunters called the mantids. Over 2,400 species species of mantids exist in the world.

Waiting for Prey

The praying mantis has a triangular head on which are perched two large compound eyes and three smaller simple ones between them. These eyes have blessed the insect with excellent vision. With its long neck, the insect can turn its head 180° to scan the surroundings.

While sitting on leaves, it is camouflaged properly because of its bright green colour. As it rests between the flowers and branches of a plant, the praying mantis waits for hours for its prey to appear. Once the prey comes closer, the praying mantis strikes with lightning speed and grabs the prey with its pincer-like spiny front legs. The rest of the four legs are used for walking.

Diet

The praying mantis is known to eat moths, crickets, and grasshoppers. Occasionally, it is known to eat very small birds such as hummingbirds and even reptiles. In Karnataka, India, a team of scientists discovered an adult praying mantis preying on small guppies, a type of fish swimming in a pond. To reach the fish, the insect used the floating leaves on the surface of the pond. Of course, the species was the giant rainforest mantis, which is said to be almost 7 centimetres in length.

▼ *A praying mantis sits on a leaf. Observe its front legs and compare it to its hindlegs*

▲ *The praying mantis has a pair of antennae with which it navigates its surroundings*

Cannibalistic Females

Female praying mantises are interesting insects. To mate, the male has to hop on the female, but if it misses the jump, it becomes the female's next meal. Also, while mating, the female might devour the male's head. The body completes mating after which the rest of the dead male is eaten by the female.

Female praying mantises then lay 12–400 eggs in groups. These are surrounded by a liquid which hardens into a protective shell. The eggs spend the winter protected inside it, to emerge in spring. Infants need the entire summer to grow into adulthood. In case of infants too, many a time their first meal consists of their siblings. Otherwise their meal is fruit flies and very tiny insects.

Isn't It Amazing!

Praying mantises are associated with various religious and medicinal beliefs. In Greek they are known as 'mantes', which means prophet. In China, it is thought that these insects help in the cure of goitre, a **thyroid** problem. The Chinese believe that eating roasted eggshells of praying mantises prevents children from wetting their beds at night. These instances reflect how the insect features in various cultures.

The Useful Bee

The honeybee is one of the most useful insects out there. Apart from giving honey and beeswax to human beings, honeybees pollinate our crops. Pollination is the process of transfer of pollen from the anther of one flower to the stigma of another flower for fertilisation. On pollination, crops produce fruits and seeds.

Threatened Bees

A queen bee lives for about two to five years on an average, but recent trends show diminished longevity too. Also, colonies of honeybees are disappearing rapidly. The European honeybee is affected by 'colony collapse disorder', a condition characterised by the sudden death of colonies. The cause for the same remains a mystery.

Changes in the world of bees can drastically affect humans as well. One example of such change is the bee orchid pollinated by honeybees. Sadly, the plant is being forced to resort to self-pollination. The anther droops forward to contact the stigma of the same flower. This is common in the bee orchids in the far northern reaches of the planet; this has happened because the local population of honeybees there has almost become extinct.

▲ *One-third of the world's agricultural crop production depends on pollination*

Human Intervention

Honeybees are already reared commercially, but now their dwindling numbers demand our urgent attention. In places where the bee population has dropped down, farmers are resorting to hiring the services of beekeepers. How is this done?

A stack of hives is delivered to a farmer by a beekeeper. The keeper then releases the honeybees on the farm. The bees head towards the plants and begin the process of pollination.

In Real Life

The Asian giant hornets are huge wasps, as big as 5 centimetres long. Their sting is said to be very painful, and their swarms can decimate a beehive in a short time, killing all the bees inside. They then feast on the pupae and larvae. Their sting is fatal even to humans.

◀ *Bumblebees scent-mark the flowers that they have visited*

Beeswax

Beeswax is used to make candles, artificial fruits and flowers, furniture and floor wax, waxed paper, as well as cosmetics.

The Bumblebee

Bumblebees, also from the bee family, are fat and furry. They are found almost all over the world, except Antarctica. These bees are excellent pollinators. A single bee can visit close to 200 flowers in one trip to eat nectar, each time picking up and leaving behind a bit of the pollen. There are two types of bumblebees—those that build nests and those that are parasitic. There are several species of bumblebees that exist within these types. The average bumblebee is 1.5–2.5 cm in length.

Isn't it Amazing!

Plants do not release their entire nectar in one go. They release it in small amounts, encouraging bees to move from plant to plant, thereby aiding better pollination.

The Builder Ants

Ants are insects that perhaps most human beings are familiar with. There are more than 12,000 species of these insects that exist today. They belong to the same insect order as wasps and bees—Hymenoptera.

The Slim One

Ants have a large head, elbowed antennae, and two sets of jaws; the front one to dig and carry food and the posterior one to chew the food. Ants are small, not more than 3 centimetres long. They come in varied colours, such as brown, red, yellow, and black.

Ants are known to 'hear' with their feet; the feet pick up the ground vibrations. Some such as driver ants have no eyes. In this case, they use their antennae to communicate. Also, they use chemical signals or pheromones to attract mates, warn others of danger, or convey the location of a food source. Ants eat nectar, seeds, fungus, and small insects.

Building a Society

Ant colonies come in varied sizes; they can range from a dozen individuals to a whopping one million of them living together. Similar to the bees, the ants show a structured society with each individual having a job based on its 'caste'.

The ant colony could have one or more queens. The queen lays eggs to create more and more members in the colony.

▲ *Ants have superhuman strength! They can carry 10–50 times their body weight!*

Builders

Ants build nests in varied ways. Some use their jaws to shovel soil around the nest to form volcano-like craters. Some build mounds which they cover with loose vegetation; these mounds help regulate the temperature within the nests.

Not all ant nests are underground. In tropical regions, some ants are seen building their nests on trees. These nests are built with paper-like substances made with soil or wood and glued together with sugar solutions produced by ants.

Usefulness

Ants play an important role in recycling nutrients in an ecosystem. They do so by feeding on and breaking down small animals, other invertebrates, and plants. In the process of building homes, they 'till' the soil, which means they bring up the deeper layers of soil to the surface. But ants can also cause damage. If their population is left unchecked, they can also destroy plants.

▼ *Ants can be found on every single continent except Antarctica*

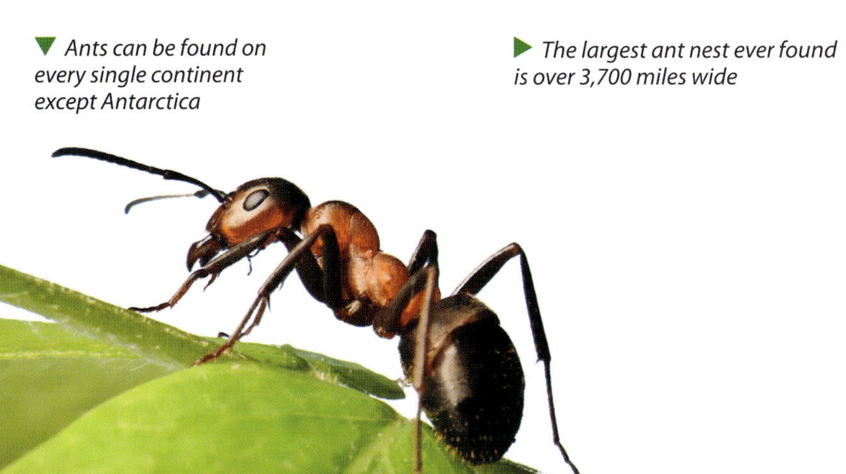

▶ *The largest ant nest ever found is over 3,700 miles wide*

Evolution of Invertebrates

The story of Earth began 4.5 billion years ago. That is a time beyond imagination. The planet was not as we see it today. It was uninhabitable due to high temperatures, volcanic eruptions, and collisions with other objects from the newly formed solar system. As time passed by, the planet stabilised. In the course of events, it gave rise to life.

First Life Forms

The first definite fossil evidence of life dates back to 3.5 billion years ago. Life originated in an atmosphere rich in methane and carbon dioxide, but very little oxygen. The first forms were **single-celled bacteria**. Fast forward to 2.3 billion years ago and a new type of bacteria evolved, called the cyanobacteria. It was unique as it produced oxygen through a process known as **photosynthesis**.

▶ *The fossils of photosynthetic bacteria found in ancient rocks are called Stromatolites*

▲ *The chlorophyll-rich cyanobacteria coat the river with green colour. These are tough creatures, still surviving on Earth in large numbers*

Snowball Earth

Photosynthesising bacteria or cyanobacteria flourished and created more and more oxygen. This changed the atmosphere, making it rich in oxygen; while most of the earlier bacteria that depended on methane died out. Then, around 715–660 million years ago, Earth cooled down to such an extent that the temperatures at the equator dropped down to -20° C. This event is called 'Snowball Earth' and is thought to have been triggered by the rapid weathering of continents, which sucked out atmospheric carbon dioxide. Life could not sustain on this snowball. It was wiped out, except for the areas near deep-sea volcanic vents.

Cambrian Explosion

About 560 million years ago, the conditions changed and the environment became more habitable—a warm Earth, rising sea levels and of course, generous oxygen levels (a result of the cyanobacteria), lead to the creation of many new life forms. This time period is called the Cambrian Explosion. It was here that the first invertebrates appeared on Earth.

Trilobites

Trilobites are the icons of the Cambrian seas. As the name suggests, trilobites had three segments. They had tough, plated bodies to protect themselves from predators in the seas, and could grow as much as two feet in length. More than 17,000 species survived for millions of years, only to be wiped out at the end of the Permian Period, 251 million years ago. A few of these species were predators and some were scavengers. Many of them ate plankton. All trilobites had antennae and legs.

▲ *Trilobites are among the most prolific fossils found in the seas*

Isn't It Amazing!

Opabinia, belonging to an extinct group of animals called *Opabinia regalis*, swam the Cambrian seas. This creature had five eyes and caught its prey with a flexible claw-like arm that jutted out of its head. It stayed close to the seafloor, hunting ancient sponges for food.

▲ *A fossil specimen of the ancient invertebrate Opabinia regalis*

Ancient Creatures

Following the Cambrian explosion, the seas were flooded with a variety of marine invertebrates along with trilobites. These were the graptolites, brachiopods, echinoderms, molluscs, corals and cephalopods. Many of these invertebrates were strange creatures. For example, the orthocone, a tailless, finless animal, was shaped like a long ice cream cone with tentacles. This 11-metre-long animal cut through waters hunting for prey such as the early species of fish and sea scorpions.

▲ *A creature called Cameroceras had a unique shape which helped it jet-propel through the waters*

Graptolites

Graptolites lived in colonies. They had tentacles and a **chitinous** outer covering. Most of these animals have been preserved as carbon impressions on shale, a type of a sedimentary rock.

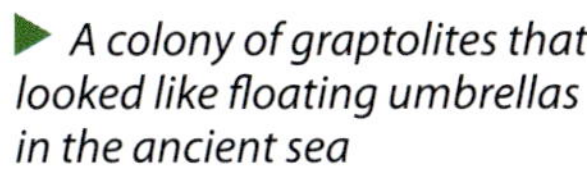

▶ *A colony of graptolites that looked like floating umbrellas in the ancient sea*

The Pterygotus

Pterygotus, a distant relative of the modern horseshoe crab, was a giant sea scorpion with some species measuring the length of almost eight feet. They were fearsome predators, holding onto their prey, such as early vertebrates as well as other sea dwellers, with their huge pincers. Apart from the pterygotus, almost 200 species of these extinct invertebrates have been identified. The fossils of these animals have been found in brackish and freshwater.

▲ *An artist's rendition of the pterygotus*

Devonian Period

Plants evolved much earlier than first thought by scientists. By the Devonian Period, which was about 416 million years ago, terrestrial vegetation started to spread. These plants did not have roots and shoots. They did not grow more than a few centimetres tall. Dwelling amongst these early plants were mainly arthropods in the form of insects, mites, and **myriapods**. So, the first insects had already evolved a few million years before the Devonian Period began. The early insects were small, wingless, and had simple antennae.

The shallow waters of the Devonian Period saw the development of large coral reefs. The seas continued to be populated by early invertebrates. However, by the end of the Devonian Period, most trilobite species had disappeared.

Passage of Time

Many large invertebrates like the *Arthropleura* grew to their size because there was more oxygen in the environment and fewer vertebrate predators on land. Eventually, these giant invertebrates reduced in size and modern arthropods came into being. It is their adaptability to the changing environment that has made invertebrates a lasting group of animals on this planet. These wonderful creatures were there before us and, according to scientists, will remain even after human beings have disappeared from Earth.

Carboniferous Period

About 360 million years ago during the Carboniferous Period, because of the growth of vast swamp forests, there was a tremendous rise in the levels of oxygen in the atmosphere. This rich oxygen-fed air allowed the arthropods to grow to humungous sizes. Arthropleura, for example, grew to almost 2 metres in length and close to 46 centimetres in width. One of its species is considered to be the largest terrestrial invertebrate ever.

▲ *The body of the Arthropleura had 30 segments like many modern millipedes*

Tough Invertebrates

According to theories, invertebrates should have been the weakest of all organisms living on Earth. They lack backbones, only a few amongst them have a proper digestive system and many, such as sponges, do not even have brains and the associated intelligence. But in reality, invertebrates are among the toughest organisms on the planet. Read on to find out why.

Unchanged Invertebrates

In the course of evolution, some species of invertebrates have remained unchanged for millions of years and are surviving happily. These are the dragonflies and horseshoe crabs. However, fruit flies are still undergoing evolution. These flies are known to make simple changes in their genetic make-up over time. These changes help them survive changing environments.

▲ *Fruit flies have adapted to survive over time*

▲ *The horseshoe crab has resisted evolution. They are seen on seashores. If the coming waves flip them over, they use their long tails to flip back*

Evolution of Resistance

The evolution of resistant, encapsulated dormant forms of invertebrates has enabled many among their species to survive in extreme conditions. For example, the eggs of freshwater fairy shrimp can remain dormant for months in dried mud, hatching only when the mud is submerged in the water again. These eggs are so resistant that even when exposed to temperatures as high as 99°C and as low as -190° C in laboratories, they have remained completely viable.

▲ *Fairy shrimp have egg sacs near their tails*

Breathing Methods

Earthworms need oxygen for survival, akin to human beings, but they have no lungs to gulp in air like human beings. That is why they breathe through their moist skin. But how do they maintain this moist skin? Not only do they live in damp soil, but their skin is also covered with a thin cuticle and slimy mucus.

Some spiders and scorpions have an interesting apparatus through which they breathe. They are called book lungs, which are a series of membranes that resemble the pages of a book. In between these membranes are air sacs which allow the air to circulate. Aiding the breathing process is the haemolymph, which is similar to the blood of human beings. For many insects, the trachea is the most important respiratory organ. This is usually made up of branching tubes. The trachea sends oxygen to all the tissues in the insect's body and takes away the carbon dioxide. The trachea is modified in insects that have to spend some time underwater. This allows them to exchange gases while they are underwater. These insects are called bubble breathers and the water beetle is the best example.

Isn't It Amazing!

Haemolymph, a fluid equivalent to blood in most invertebrates consists of necessary nutrients which help the organism survive, but insects lack red blood cells as well as haemoglobin, giving haemolymph an almost colourless appearance.

Comb Jelly

The freely swimming comb jelly maintains its spherical shape through the presence of water in the internal canals. These canals support eight rows of comb plates, with which the comb jelly swims. Each comb plate is covered with hair-like cilia which propel the organism forward. The organism swims mouth-first.

▶ *Comb jellies are 95 per cent water*

Structure

The stony coral polyp obtains support from a mineralised theca or a cup which is secreted by the animal. This helps anchor it to the basal structure of the coral colony as well as to its neighbours.

The closely related sea anemone does not use a rigid structure like the stone coral. Instead, it supports itself by using the water circulating around its central cavity. The water is drawn in through grooves on the sides of the cavity and expelled up the centre. It helps the organism maintain shape and internal pressure.

▼ *Hermit crabs have asymmetrical bodies that curl towards their right side*

Protective Armour

Invertebrates that are arthropods or joint-legged animals, such as scorpions, crabs, lobsters, and spiders have hard, chitinous, water-resistant exoskeletons which protect them from dehydration and various environmental changes such as extreme heat or heavy rains, as well as predators. This exoskeleton is moulted regularly by the invertebrate; it is while the exoskeleton is growing back that the invertebrate, with its exposed softness, is most vulnerable.

Molluscs such as snails have developed shells for protection, while hermit crabs live in discarded sea shells that wash up the shore, moving to a bigger one as they grow. But what is interesting is that this crab always finds an empty shell; it never kills or throws out the occupants of the shell.

Symmetry

Symmetry is the proportionality or evenness that is displayed on animal bodies. Most invertebrates are divided into three types of symmetries, namely bilateral symmetry, radial symmetry, and asymmetry. In bilateral symmetry, the body is divided into two equal parts by an imaginary line. Beetles, crabs, and lobsters have bilateral symmetry. In the case of radial symmetry, the body is oriented in a way that an imaginary line radiates through its centre, like in animals such as jellyfish and starfish. Asymmetrical organisms, such as sponges, are those with no symmetry; the body parts in such organisms do not correspond with each other to create a defined shape.

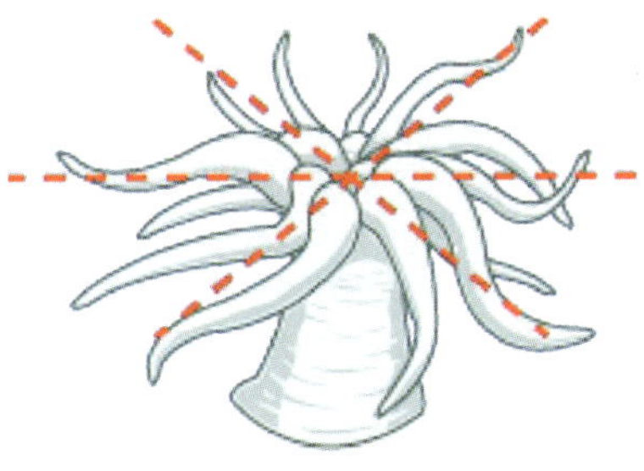

▲ *A coral polyp has radial symmetry*

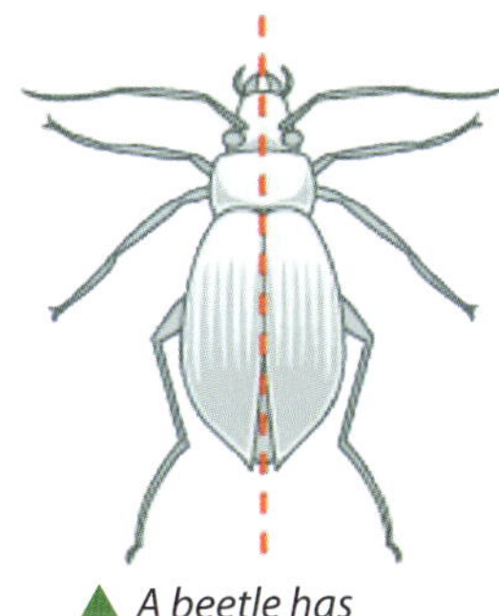

▲ *A beetle has bilateral symmetry*

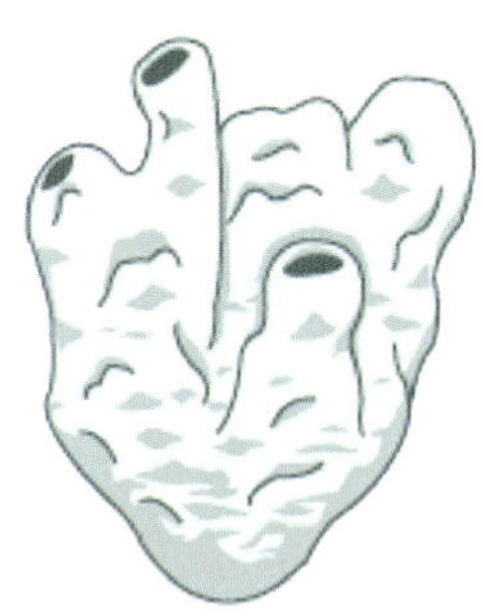

▲ *A sponge has no symmetry*

Chemical Warfare

Invertebrates have one of the best chemical defences in the animal world. They use an array of these defences, ranging from jets of boiling caustic fluids, paralysing toxins, and foul-smelling liquids to potent poisons that can cause immense harm. Chemical defence is commonly accompanied by a distinctive colouration or behavioural pattern that serves as a warning to the predators.

Borrowed Threads

Sea slugs lack any form of armour like shells, claws, or spines, but they are rarely bothered by even the hungriest of fish. The secret lies in the projections that cover their bodies. Many of these projections contain cnidocytes or stinging cells, which give off a barbed, poison-tipped thread the moment anything touches them.

Sea slugs themselves do not produce this poison. They retain it from their prey, the sea anemone. When sea slugs eat the sea anemones, the stinging cells from the latter's tentacles are not digested. These are transported by the slugs to the projections on their bodies and stored to be used in defence.

▲ *The bright colours of a sea slug serve as the first warning to its predators*

Velvety Smarts

Onychophorans, also known as velvet worms, live in tropical regions. These are known to squirt an odourless fluid from the glands on the ends of the projections on their heads. It can be squirted from a distance of almost 15 centimetres. The fluid then hardens immediately, disabling the predator.

◀ *Velvet worms have been around for over 500 million years*

Whiplash

A whip scorpion raises its abdomen when it senses danger, and releases an acidic spray on its predator, which it dispels from its large glands.

◀ *Whip scorpions do not have tails. They resemble spiders*

Boxy Defence

Box jellyfish are named so because of the box-like appearance of their heads. These are transparent blue in colour. The poison of the box jellyfish is considered to be one of the deadliest in nature. It affects the heart, nervous system, as well as skin. It is fatal not just to animals but to human beings as well. If attacked, the victim undergoes intense pain and could die of heart failure. If the victim does survive, the pain can last for months.

▶ *The venom released by the box jellyfish can even stun its prey*

Clever Camouflage

Camouflage means to blend in the surroundings. Invertebrates use different methods of camouflage while waiting for a prey or while hiding from a predator. Many animals have been using this tactic for millions of years. They are impressive in their array of colours and patterns. Invertebrates might be tough, and they may possess chemical weapons, but in the big bad world, being small creatures, many need to use camouflage to survive.

The Sandy Ghost

Ghost crabs blend well in the sand dunes that are available in plenty on the beaches that they inhabit. The word 'ghost' in their name refers to the fact that these crabs can quickly disappear from sight using their six strong legs. They use camouflage well and lie in wait on the beach, with just two eyes protruding out. The moment a prey such as other crabs, lizards, insects, or clams approach them, the ghost crabs grab and devour it. To hide from the predator, ghost crabs simply vanish into their small burrows.

▲ *Ghost crabs are also known as sand crabs*

Light Play

Cuttlefish belong to the mollusc family. They have eight arms and two tentacles which are used to capture the prey. The cuttlefish have the amazing ability to change colour within seconds. They do it with the help of pigmented organs called chromatophores. The invertebrate expands or contracts these chromatophores, creating an array of colours. The colour-changing property of the cuttlefish is one of the best camouflage techniques on the planet.

▲ *Cuttlefish have venom which can be deadly for human beings*

▲ *The goldenrod crab spider often sits on goldenrod flowers or daisies which are white or yellow in colour*

Deadly Golden Touch

The goldenrod crab spider, or flower spider, can blend in with a range of flowers by changing its colour. This change of colour from white to yellow takes about three weeks; the reverse takes six days. The spider sits in wait for an innocent prey on the flower, grabs it with its front legs, and kills it with its venom.

Hidden Beauty

Oregonia gracilis is found in the north-western coasts of the USA and Japan. It is also called the decorator crab and has a graceful appearance. Using its front legs, the crab picks up algae, sponges, wood chips, and any other marine detritus. The fragments are then modified by the crab using its mouth.

▶ *Sometimes a decorator crab looks like a moving rock. It camouflages itself with different items, which is why it has been given the name 'decorator'*

Parasitic Invertebrates

Living inside another plant or animal can be a good source of nutrition. This is exactly what parasites do to have their needs met. A parasite lives inside a host without a symbiotic relationship. The host provides the parasite with food, water, and shelter. In return, the parasite might harm the host or simply live without causing major issues.

Tapeworms

Tapeworms are flat, segmented, and can parasitise the digestive tracts of many vertebrate species, including human beings. The largest of the tapeworms live in sperm whales and can grow to almost 100 feet. In human beings, tapeworms can be as long as 60 feet.

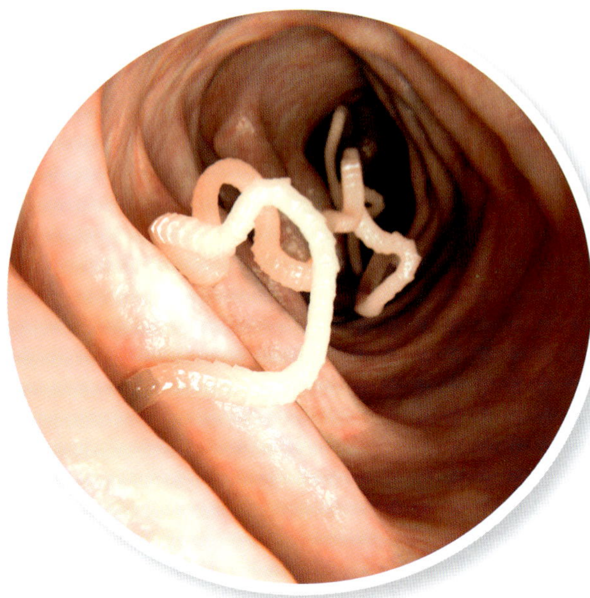

▲ *Tapeworms are caused due to the intake of contaminated food or water*

Threat to People

Most often tapeworms cause weight loss, abdominal pain, loss of appetite, diarrhoea, and weakness in human beings. In rare cases, they may lead to intestinal blockage. In this case, the doctor gives a course of medicine to clear out the infection.

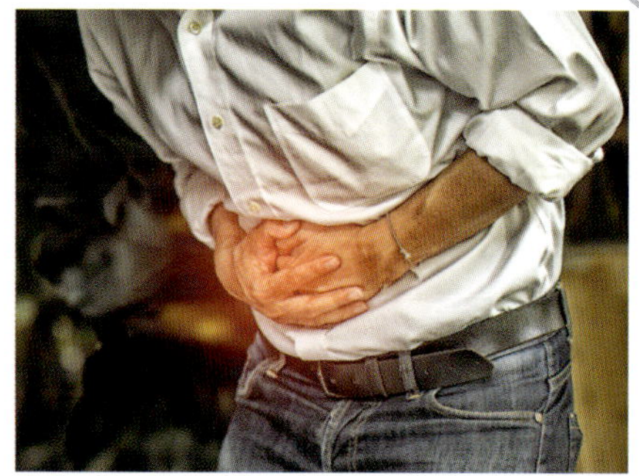

▲ *Adult tapeworms can live up to 30 years in a host*

In Real Life

Ascaris lumbricoides is a parasitic roundworm which causes infection in human beings. The infection is common in warm tropical regions. The *ascaris lumbricoides* can grow to a size of 35 centimetres! If the host carries both male and female worms, on fertilisation, a female can lay more than 2,00,000 fertilised eggs per day.

A Rounded Existence

Nematodes, also known as roundworms, are one of the most abundant animals on Earth. They have a range of habitats and can live as parasites in plants, animals and human beings. They live in soil, freshwater, marine environments and even in items such as vinegar and beer. Roundworms come in a range of sizes, right from the microscopic variety to ones that are as long as 23 feet.

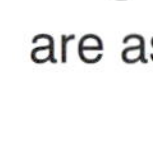

The Oak Apples

Hard, round marble galls, also called oak apples are commonly seen on oak trees. It is the tree's response to an infection caused by a parasitic insect called the gall wasp. The female lays eggs on the tissues of leaves or twigs of the oak tree. When the eggs hatch, the tissues around the larvae, also called grubs, swell up, creating gall. When the grub leaves the gall, it creates a hole. Most often, the affected twig or leaf is shed off by the tree. In general, galls do not affect the life of an oak tree, unless there are numerous ones on the bark as well.

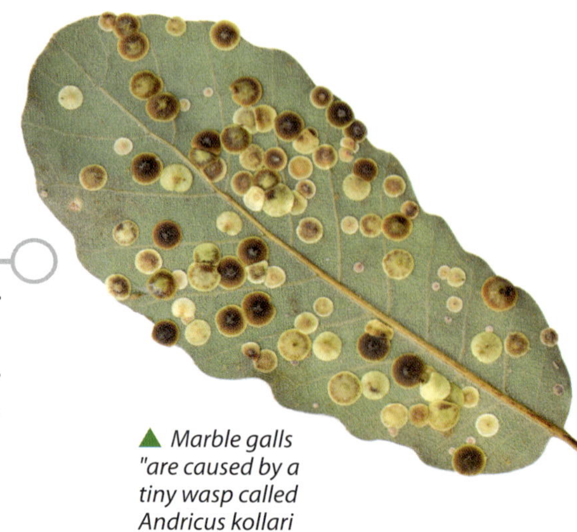

▲ *Marble galls "are caused by a tiny wasp called Andricus kollari*

Earthworms & Leeches

Worms are so varied and numerous across the world that they form different groups. They are found in oceans, burrowing in land, or even attaching themselves to the host as blood suckers. There are over 9,000 identified species of these worms in the world today. Leeches are slightly flat-looking worms. These muscular worms can lengthen and shorten their bodies as desired. They have 34 segments on their bodies along with suckers, which distinguish them from other worms.

Earthworms are also known as angleworms

Habitat

Earthworms are native to Europe. However, they are found in big numbers in North America as well as western Asia. They do not live in the desert regions or areas with permanent snow coverage as such conditions do not suit them.

Burrowed Life

The average length of the earthworm might be 7–9 centimetres, but it can burrow almost up to 6.5 feet deep into the soil. It can burrow all day, and crawl above to feed at night.

Some leeches prey on small worms, insect larvae, and snails. The horse leech can grow up to 16 centimetres long and swallow its food whole.

Diet

The mouth of the earthworm is at its first segment. They eat soil, taking in nutrients from the dead leaves and roots. A single earthworm can eat almost one-third its body weight.

The leech clamps itself onto other animals. It then uses its three sets of tiny teeth to make its way into the skin. It sucks blood and other bodily fluids up to almost five times its weight.

A leech can survive for weeks before needing another feed. This is because there is a pouch in its digestive system where it can store the food that it eats for weeks or even months.

Earthworms do not have eyes, but they can sense light

Reproduction

Earthworms are hermaphrodites, which means a single animal will have both male and female parts. But they do not self-fertilise. They need another earthworm to create the next generation.

Usefulness

Leeches can suck blood from their host and make it weak. They also attach themselves to human beings. The use of leeches in therapy or treatment is called 'leech therapy'. This was very common before modern medical practices evolved. Doctors used leeches to increase blood circulation and break down blood clots.

Each leech has: 10 eyes, 6 hearts, 10 pouches for storing blood, 32 brains, and 200 enzymes

Crabs & Other Crustaceans

Crustaceans belong to the arthropod group. Crabs, lobsters, shrimps, and prawns are common examples of crustaceans. They have tough, calcareous exoskeletons. They also have two pairs of appendages in front of the mouth and paired appendages near the mouth which function as jaws. There exist about 45,000 species of crustaceans across the world. Most of these are aquatic in nature.

Crabby Behaviour

When we talk about crustaceans, the most common one that comes to mind is the crab. Crabs are a delicacy around the world. A typical crab has a hard, shell-like cover with a small abdomen tucked inside it. It has five pairs of limbs. One pair consists of large pincers and four pairs consist of the limbs used for walking. Crabs live in sea water, but there are exceptions which live in rivers and lakes.

▲ *The stone crab can regenerate its claws if they are removed*

Survival

The crab is an omnivore, which means it eats both plants as well as meat. It eats mussels, smaller blue crabs, snails, fish, and even carrion.

Large male crabs can have a shell width of almost 22 centimetres. These crabs are excellent swimmers because they use their paddle-shaped hind appendages to swim. Unfortunately, since they are sensitive to the environment and climatic changes, their numbers are on the decline.

Cousin of the Crab

Lobsters, like their crab cousins, live in varied habitats such as oceans, brackish waters, and freshwaters. Like crabs, lobsters have five pairs of appendages. While four pairs allow them to walk around, the remaining one forms the sharp pincers used to hold and crush the prey. The four walking leg pairs are attached to the thorax, which is the middle part of the lobster's body. The abdomen in the lobster is at the rear end, like a tail. This invertebrate is well-protected with a tough shell. The abdomen is covered with a series of plates. Lobsters are known to moult in order to grow.

Lobsters cannot see properly, but they have a sharp sense of smell and taste. They use their long antennae for the same. They feed on shellfish, small fish, algae, and plankton.

▲ *Most lobsters are nocturnal, scavenging marine animals. There are many species of lobsters; some of them are called true lobsters.*

Woodlice

Woodlice look like insects but are crustaceans. There are about 3,500 species of woodlice in the world and these are among the few crustaceans that live on land. These invertebrates have 14 legs and a thick exoskeleton. When they grow too big for the exoskeleton, they moult.

Woodlice have two antennae at the front end with which they sense their surroundings and two small outgrowths called uropods at the rear, which they need for navigation. Uropods also produce a chemical in some woodlice species. This chemical is used to ward off predators. Woodlice survive off of rotting plants, fungi, and their own excretory matter. They live in dark and damp places, on walls, and on decaying tree barks.

Creepy Crawlies

Spiders belong to an arthropod group called arachnids. The main feature of this group is that the animals here have 8 legs unlike crustaceans, which have 10 or more, while insects have 6 legs. Arachnids do not have wings or antennae. There are close to 80,000 species of arachnids spread across the world in a range of habitats, out of which more than 45,000 are spiders.

Appearance

The body of the spider is divided into two parts. The first part is the front cephalothorax, which consists of the spider's eyes, stomach, brain, glands that make poison, and mouth fangs. The second part is chelicerae or the spider's muscular jaws. It is the chelicerae that hold onto the prey while the spider injects it with poison.

Spiders cannot chew their food. Their fangs are like straws. They deliver enzymes into the prey, which make the prey mushy like a soup. Then they suck up the prey.

The spider also has leg-like pedipalps. These are not used for walking, but more as antennae to sense the objects that come in its way. It needs the pedipalps to hold its prey and even spin the legendary spiderweb.

The back of the body is called the abdomen, which contains important internal organs. It is at the end of it that the silk-producing glands called spinnerets are present.

▲ *Spiders have four pairs of eyes and, unlike other arthropods, have very keen vision*

Spiders have a lot of hair on their legs. It is this hair which picks up vibrations and scents from the surroundings. Similar to other invertebrates, spiders have tough exoskeletons, which do not grow with them.

Webs

To spin a web, the spider needs some support like an anchor for the web's threads. This support could be rocks, vegetation, or any other solid surface. The spider selects a high perch and lets the thread travel until it touches and sticks to another object. More threads are spun, making a dense network.

Most webs are traps, but do not trap the spider itself. This is because it walks only on dry threads and uses special brushes on its claws to grip the threads. Spiders also release oil which forms a coating on their feet to prevent them from sticking to the web.

The moment the prey touches the web, the silk wraps around it. The spiders release the poison using their fangs. Once the prey is paralysed, it is unwrapped and consumed.

▲ *The underside of a spider has thousands of silk glands that extrude silk through spinnerets*

Poisonous Spiders

Though almost all spiders have poison, not all are harmful enough to hurt human beings. But there exist a few species of spiders with poison so powerful that they can harm most animals, including human beings. The black widow spider is a fierce arachnid, with its poison said to be deadlier than that of a rattlesnake. Its sting affects people too. If a human being is injected with the poison, they might have nausea, aches in their muscles, and breathing problems. The female black widow spider, after mating, at times eats the male spider. A close cousin, the brown widow spider, has poison twice as potent, but it is less aggressive.

▲ *Black widow spider; widow spider nymphs are mostly white when they hatch from the egg sac*

What Makes a Mammal?

Mammals are classified as 'Mammalia'. This class of animals have a few unique features that make them the highest order of beings residing on the planet. Besides giving birth to their young and feeding them milk from their own bodies, there are other features that determine whether an animal species is a mammal.

▲ *Naked mole rats cannot control their body temperatures very well, so they like to live in tunnels to keep cool*

Hair

Hair refers to small thread-like projections that emerge from the outermost layer of the mammal's skin. It is also called the fur or animal coat. All mammals possess this trait, though they may have hair in varied appearances and at different stages of their lives.

Dolphins have small pits at the snout, where they had hair at the time of their birth. Dolphin calves in their mother's womb have these pits filled with hair, but after birth, the hairs fall off while swimming. Animals like elephants and rhinoceroses have sparse, bristly hair, while polar bears have a long, thick fur to keep warm in the cold Arctic. Human beings are among the mammals that have lesser hair.

▲ *Polar bears need their fur to keep warm in freezing habitats*

▶ *A lioness with her cub; according to experts, there are only about 20,000 of these fierce cats left in the wild*

Warm Blood

Mammals are warm-blooded animals. In other words, unlike amphibians or fish, they can maintain their body temperature, regardless of being out in the snow or sunlight. Their average body temperature varies between 36.7° C and 37.2° C. This ability of mammals to produce heat within their own bodies is called **endothermy**.

How do mammals maintain their body temperature? They might shiver. This action increases the heat that their bodies produce. To keep cool, they might pant or sweat, which leads to heat loss. Some mammals might hibernate. This reduces the heat in the body as other bodily functions slow down.

Giving Birth

Majority of mammals (with the exceptions of the duck-billed platypus and echidna) do not lay eggs. The mother bears the young ones inside her body in an organ called the uterus until birth. Such organisms are called **viviparous**. Except for **marsupials**, who carry the young ones in a pouch, mammals give birth to a well-developed baby. The time the **foetus** spends in the womb of the mother is called the gestation period, which differs in all animals. An elephant carries its baby for almost 22 months, while the Virginia opossum carries its baby for only 12–13 days.

Mammalian Milk

Once the baby is born, it is fed milk produced by a special organ called mammary glands. This unique characteristic defines mammals. Non-functional glands are seen in males too, but it is the females which carry glands that can produce milk to feed the baby.

In primitive animals like the duck-billed platypus, the glands express milk directly onto the skin or fur, which is then lapped up by the baby. Mammary glands are modifications of sweat glands, the organ that produces sweat.

Isn't It Amazing!

Humpback whales are very agile for their size. How does the huge animal manage to be so swift? That is because it has golf-ball sized bumps on its head and **pectoral fins**. They are **hair follicles** connected to sensitive nerves. Studies reveal it is these that help the whale in being swift. Also, the hair follicles are said to help the animal detect water currents and food.

The Latin name for humpback whales is 'Megaptera novaeangliae', meaning 'big wing of New England'

Bats have extraordinary hearing abilities

More Mammalian Features

In all other forms of **vertebrates**, the jaw is attached to the skull via a set of bones. However, in mammals, early on during evolution, the jaw hinged itself directly to the skull, to create a set of ear bones; this helps this class of organisms hear better. Also, mammals have different types of teeth, such as the molars, premolars, canines, and incisors. These specialised teeth help them grind and chew their food. The biggest evolutionary development in mammals is their big brain, which helps them think and reason better.

In Real Life

Elephants can overheat easily because they do not have a large skin surface in comparison to their body size. These animals do not sweat like human beings; instead, they use their skin to keep cool. If the skin heats up on a hot day or after a long walk, the internal heat generated cannot be reduced. This is where their hair comes to the rescue. It helps the body stay cool!

Small, sharp hair seen on an elephant's head

A Blast from the Past

Though it might be hard to imagine, it took a long time for the different species of mammals to evolve into what we see today. If we want to read about the evolution of different mammals, we can do so because of the researchers and scientists who have unearthed fossils or remains of extinct mammals that existed long before the advent of human beings. Through the study of these fossils, scientists learned about the existence, behaviour, food habits, and extinction of ancient mammals.

Sabre-toothed Cats

If we were alive 42 million to 11,000 years ago, then we might have been able to spot sabre-toothed cats in the wild. Sadly, these mammals are extinct. They were said to live in North America and Europe as their fossils were found around these continents. There were different species of sabre-toothed cats and some of them prowled the wilds of South America too.

Sabre-toothed cats had short, bulky bodies and short tails when compared to modern lions. These vicious cats were known for their sabre-like teeth, which were very long and sharp. They were named after this feature.

▲ *It is misleading to call these animals sabre-toothed tigers as they are not closely related to modern tigers*

▼ *The Deinotherium is also called the 'Hoe-tusker'*

Deinotherium

The name *Deinotherium* means 'terrible beast'. At a glance, this large ancient mammal resembles the modern elephant. But compare them closely and their differences become clear. The *Deinotherium* had a shorter trunk and downward-facing tusks. Scientists believe they were used to pull down trees or dig out roots from underground. It was also much larger than the modern elephant, nearly 13 feet tall. It lived in Europe, Asia, and Africa. The African species outlasted the other two. The mammal disappeared around 7 million years ago.

Isn't It Amazing!

Even though they are named for their famous teeth, it is believed that at least two species of the sabre-toothed cats had weak jaws and bites. These mammals did not chase their prey. Instead, they ambushed it with their packs. They bit the prey but primarily used their neck muscles and forelimbs to hold it down and weaken it. Research has revealed that modern lions might have stronger bites than these ancient beasts!

Glyptodont

Think of the armadillo. Now, picture a creature that is much larger in size and weighs nearly 2,000 kilograms. It is the glyptodont. Compared to that, the modern armadillo (a distant cousin) only weighs up to 54 kilograms!

The glyptodont originated about 20 million years ago. It had a powerful spiky tail that was shaped like a club. It had a tough **carapace**—the bony outer shell covered with scales. It lived in South America and is said to have disappeared at the end of the last Ice Age, about 10,000 years ago.

▶ *The extinction of glyptodonts coincided with the arrival of early humans in the Americas*

Entelodont

The extinct entelodonts and modern pigs share a long, extended family. They might even have been related to the modern hippopotamuses. Fossil evidence shows that the entelodonts originated about 40 million years ago and became extinct 16–19 million years ago. These ancient mammals roamed in Asia, Europe, and North America.

The animals were medium or large with the heaviest species found to be nearly 900 kilograms. Naturally, they were much heavier than the modern pigs. The entelodonts probably had an **omnivorous** diet.

◀ *The entelodont had a fierce temper and was extremely fast*

In Real Life

Along with the sabre-toothed cat, the woolly mammoths make the movie *Ice Age* interesting. In reality, they were as big as African elephants but had small ears, probably to keep out the cold of the Ice Age. They had very long tusks and were herbivores. They disappeared when the climate turned warmer.

▲ *As the name suggests, the woolly mammoth's body was covered with woolly hair to keep warm*

The Walking Whale

In 1992, a **paleontologist** named Hans Thewissen discovered the first fossil of a whale in Pakistan which was, at the time, about 49 million years old. This fossil was of an animal called *Ambulocetus natans*, which had four legs. They were of an older ancestor or relative of the whale called *Pakicetus*.

On examining the jaw and ear, Thewissen discovered that the fossil resembled the structure of a modern whale and had strong hind limbs. He believed that the whale's ancestors began to evolve into their aquatic lifestyle nearly 50 million years ago.

Incredible Individuals

Hans Thewissen is a Dutch paleontologist and a researcher in the field of anatomy. He focuses on the study of whale fossils to understand how they evolved from land animals. He is also involved in the conservation of beluga whales and does fieldwork in Asia, Alaska, and the neighbouring regions.

Our Evolution

Do you ever wonder how human beings evolved from apes? Going by the knowledge gained so far, we have discovered that human beings are the most skilled and developed mammals in the history of the evolution of Earth. The story of humans started around six million years ago in Africa. Our primate ancestors spent most of their time on trees, just like modern primates. So, what changed?

The Rough Details

It is said that an ancient ancestor of human beings came down from the safety of the trees and began to roam the lands—it is not yet certain if they roamed grasslands or forests. It was probably at this point that humans became different from our closest living relatives, the chimpanzees and other such apes.

Scientists have put the human tribe into a group called 'Hominin'. The modern human beings (called *homo sapiens*) are the only living members of this tribe, but fossil evidence shows that many hominins preceded us. The history spans more than a million years:

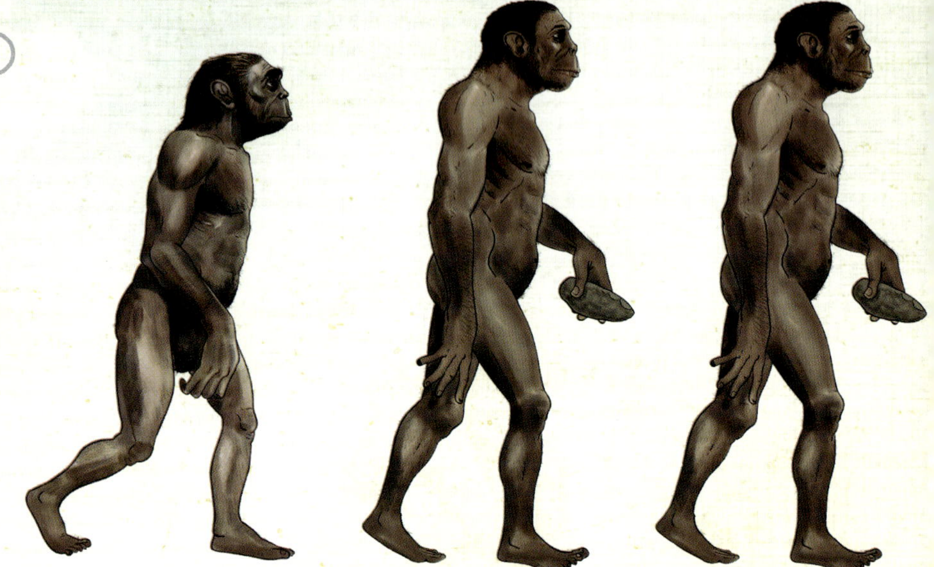

4.4 Million Years Ago

Ardipithecus, so far, is considered the first hominin. While there are ape fossils older than *Ardipithecus,* there is no evidence of them being human ancestors. *Ardipithecus* appeared in forests of present-day Ethiopia. It was probably not a good walker or runner, but its fossils indicate the presence of the same anchors for muscles we have on our pelvis, which chimpanzees and other apes lack.

3 Million Years Ago

Our first true human ancestor, *Australopithecus* appeared in Africa. A female *Australopithecus* was discovered in 1974 in Hadar, Ethiopia. She was named 'Lucy' after a song by the Beatles. The fossil evidence shows that Lucy was a bipedal, which means she walked on two legs. She was seen to have wisdom teeth. *Australopithecus* had a larger and a more complex brain structure.

2 Million Years Ago

Almost 2-million-year-old fossils of *Homo habilis* were discovered in Koobi Fora in Kenya and Olduvai Gorge in Tanzania. The word means 'handy man' or 'hand-using man'. The fossils show that the brain had increased in size and complexity yet again. Scientists feel that *Homo habilis* could use stones as tools. It was the development of such skills that set humans apart in the animal kingdom.

1.7 Million Years Ago

Homo erectus, which refers to the species' relatively erect posture, was the first ancestor to have human-like body proportions. This species first appeared in East Africa and spread to various parts of Africa, Asia, and Europe. Along with the use of stone tools, these early humans discovered fire.

Isn't It Amazing!

Did you know humans are mammals with the longest hair? If you did not cut it, your hair would reach the floor and beyond. Beards and moustaches can also grow very long. A close second is the musk oxen, an animal which lives in the Arctic. It has a long, brown coat which reaches almost down to its feet. The Arctic fox has a thick coat of fur which can survive temperatures that drop below -50° C. Its coat turns from white to brown during the cold harsh winter and even grows longer for more protection.

History of humankind (L–R: Ardipithecus, Australopithecus, Homo habilis, Home erectus, Homo neanderthalensis, Homo sapiens)

400,000 Years Ago

Around the time *Homo erectus* was roaming the various continents, a new type of hominin appeared in Europe. They were called *Homo neanderthalensis* or simply Neanderthals, named after the Neander Valley in Germany, where the first fossil was discovered. These are the closest relatives of mankind on the evolutionary map. Their bodies were shorter, and they had larger brains. They also had angular cheek bones and large noses.

Community Living

Neanderthals were the first to exhibit the traits of civilisation. They controlled fire, built shelters, and used stone tools for hunting. They practiced burial rituals for the dead. They were the first to develop art in the form of cave paintings which show their culture, society, and the world they lived in. They disappeared nearly 40,000 years ago. They either assimilated with modern *Homo sapiens* or became extinct.

200,000–100,000 Years Ago

Homo sapiens are thought to have evolved in Africa. The oldest remains unearthed so far are from Morocco. All of us living on planet Earth today belong to this species. *Homo sapiens* have the most complex and largest brain capacities amongst any hominin species. They have an erect posture, light jaws, and small teeth.

Development of Society

Human beings, as they are called, colonised and spread across the different continents on Earth. Our ancestors began the use of complex tools, not just for hunting, but farming as well. They built villages, which created societies. They created art, music, foods, ornaments, clothes, and rituals, which form the basis of human civilisation as we know today.

The Killer Whale

Though they live in the water, the killer whales, or orcas, are some of the most fierce and predatory animals on the Earth. With a large, powerful build that supports a top speed of 56 kmph, and a smart, cunning mind, the killer whale is very well named.

Meet the Killer Whale

Killer whales are also called orcas. Do not be confused by the name though, because they are technically dolphins. These giant carnivores are around 20–26 feet in size and 6,000–10,000 kilograms in weight. They are so large and fierce that they do not have a natural predator. The biggest threat to their existence is human beings.

▲ *Orcas stick to family groups called 'pods'*

Appearance

Imagine that you have lost a tooth, but it does not grow back again. That is what happens to killer whales. They only have one set of teeth which are large and interlocking. Killer whales have 40–56 teeth; each is about 7.6 centimetres long. They are sharp and can tear apart the prey easily.

The black and white skin of killer whales is constantly renewed. Therefore, they have very smooth skin that helps them swim fast. They have 7.6-10 centimetre thick layer of blubber beneath their skin.

Diet

Killer whales do not chew their food, they swallow it whole. They eat sea lions, small seals, penguins, walruses, squids, sharks, and fish. Orcas eat about 230 kilograms of food daily.

Predatory Nature

Killer whales do not have the playful nature that their cousins—the dolphins—are famous for. They might be smart and social like dolphins, but they are also dangerous—one of the top predators of the sea. They live in large pods or family groups. All members of the pod participate in hunting. They hunt using strategies very similar to wolves, such as encircling the prey before moving in for the kill.

Killer whales have black backs, but their stomachs are white. So, a seal sitting on an ice floe might not see the whale, because the black blends well with the water.

They sneak up on and bump off the seal from the floe, into the water, to catch it. Similarly, a fish might not see the white underside as it blends with the light streaming down in the sea from the surface.

◄ *Killer whales also eat krill to maintain their body weight*

All About Marsupials

Similar to other mammals, marsupials have hair and mammary glands. Over 250 species of marsupials exist. They are characterised by the continued development and maternal dependency of their newborns.

Young Marsupials

The young are very vulnerable at the time of their birth and are heavily reliant on the mother's milk. They fasten their mouth firmly on the mother's swollen up teats, and continue to develop for months in her marsupium. The marsupium or the pouch is a flap of skin that covers the mother's lower abdomen and teats. After a few months of warmth and shelter in the mother's pouch, the baby is gradually weaned off.

Baby kangaroos are known as joeys

The Anteater

If you see an anteater, you will wonder how this toothless, strange-looking animal functions. They have a very long, worm-like tongue that helps them gulp down termites and ants easily. They have a tube-like snout and powerful, big claws that can tear down strong anthills. Their eyes and ears are small and rounded and they have very poor eyesight and hearing. Their brain is round and small too. But it is interesting how the anteaters' sense of smell is considered to be 40 times stronger than that of human beings! They live in tropical grasslands and forests in Central and South America. Anteaters easily eat a few thousand ants in one sitting and then abandon the nest. The biggest species are over 40 kilograms.

The biggest species of anteaters is the giant anteater

Isn't It Amazing!

The modern kangaroo is a herbivore. Can you imagine a kangaroo that eats flesh? There was an extinct marsupial called *Ekaltadeta* closely related to the cute, jumpy kangaroos of today. They are also known as the killer kangaroo. As big as the modern dog, these killer kangaroos had large teeth and held down their prey with their front legs.

Koala

Koalas, also referred to as koala bears, are not bears, but are marsupials. They live in Australia on eucalyptus trees. They can eat more than a pound of leaves in a day. Mind you, they are fussy eaters and only choose the best leaves. They do not drink much water as they get all the water they need from the leaves they eat. The animals sleep more than 18 hours a day. A koala baby, when born is called the joey, and is blind and earless.

Kangaroos

Kangaroos are large animals that hop from place to place, at speeds as high as 56 kmph. They travel far and wide in search of food and water. These mammals have a pouch in which they carry their joey and belong to the *Macropodidae* family of marsupials that includes the happy quokkas and wallaroos.

Koalas eat up to 1 kg of eucalyptus leaves every day

The Intelligent Primates

Scientists are known to have studied the teeth of primates to make findings about them. Around 57 million years ago, there lived a mammal which was probably the first true primate to have ever lived. It was a small animal named *'Altiatlasius'*. Its fossil was discovered in Morocco. Scientists feel it could be an important link in the evolution of primates.

What are Primates?

A primate is mammalian group which includes monkeys, lemurs, tarsiers, apes, and human beings. Primates have big brains and are intelligent animals. Their eyes look straight ahead from their face.

Each of their hands has five fingers and each hind limb or foot has five toes. Primates are dexterous, meaning they can use their hands skilfully. Some ape species and human beings even have opposable thumbs.

▲ *Some trainers have managed to train chimpanzees to perform sign language as part of an experiment*

Close Cousins

Chimpanzees are probably the closest relatives of human beings in the primate family, apart from bonobos. These primate species reside in the tropical forests of Africa and are omnivores. They are extremely intelligent, with logical thinking and problem-solving capabilities. Chimpanzees live in groups called communities and are extroverted by nature. They are known to display love and affection towards others. The mother and child, and siblings share strong bonds and, just like human beings, even they can display a bad temper.

Orangutans

'Orangutan' in Malay means 'person of the forest'. This primate is long-haired, orangish in colour, and has an enormous arm span (almost 7 feet in males) when stretched. Orangutans are found in Sumatra and Borneo.

An average orangutan spends most of its life in trees. It makes itself a bed in the tree by breaking apart branches, twigs, and leaves. It makes a new bed every night, and uses large leaves as an umbrella to keep itself dry from the constant tropical rains.

▶ *Orangutans are semi-solitary creatures, which means they like to live around two or three others at a time*

Sun Soakers

Found in Madagascar and a few neighbouring islands, the ring-tailed lemurs are great sunbathers. Every morning, they sit on the ground with their arms open wide, facing the Sun. These animals live in groups of 15–20.

A ring-tailed lemur has a long, striped, black and white tail. The animals mainly eat fruits. They have powerful scent glands, which they use to mark their territory, communicate, and attack predators.

◀ *Ring-tailed lemurs look for food together. After eating, they like to relax in spots that receive sunlight*

Shy Primates

Tarsiers, with their spooky eyes, are said to bring bad luck by villagers in Borneo. They frighten many with their appearance. These small monkeys are residents of Southeast Asia. Their striking feature is their remarkably huge eyes that they use to see better at night, as they are nocturnal animals.

▶ *Tarsier babies are the largest as compared to their mothers*

Conservation of Endangered Mammals

We share our planet with a variety of flora and fauna, and all species have an equal right to call it home. Apart from that, they make the world we live in beautiful and interesting. They provide for many of our needs; for example we get wool from sheep and milk from cows. Upon visiting a nature park, one thing we all look for are animals. So, is it not our duty to contribute toward saving the animals?

▲ *Many animals form large herds for migration*

Ways to Help

- Educate people about the importance of animals.
- Make the world in which animals live a better place for them to inhabit. It could be done by cleaning up human waste from their surroundings—removing plastic and other harmful waste thrown around in parks and beaches.
- Volunteer at a zoo, aquarium, or natural reserve in your area.

Let us look at a few animals from around the world which need our help. We may not have seen them if they live in faraway lands. But every animal in this world counts and if they are not conserved, the vast diversity of animals we see might be lost forever.

▲ *The Sahara Conservation Fund is active in protecting the addax*

Lost in the Desert

Addax is a desert antelope known to live in harsh environments. These antelopes are active during nights, since in the daytime they dig into the sand and stay under cover to keep away from the hot desert sun. Once found in abundance around Sahara in Africa, they now live in a confined area between Niger and Chad.

Sadly, this animal is in the Critically Endangered Red List of IUCN. A protected population exists in the Yotvata Hai-Bar Nature Reserve in Israel. Soon, Niger will have a vast area reserved to protect the remaining animals that move in the wild.

The dwindling number of the addax is mainly due to hunting. This mammal's meat and leather are useful commodities purchased by the local people. Droughts and a shrinking habitat also contribute to their endangerment.

The Spotted Predator

The Asiatic cheetah has long legs, a spotted coat, and a slender frame. It is one of the fastest runners on land. Once upon a time, these animals abounded the Indian sub-continent, Afghanistan, and Central Asia. However, now they fall under the Critically Endangered list of the IUCN. The animals prefer to live in plains and desert-like areas.

Today, their population is concentrated in a remote region of Iran. This reduced population was caused by the loss of prey and hunting activity. The government of Iran is working with various wildlife agencies to create a programme which might save this animal in one of its last reserves.

▼ *There are only about 50 Asiatic cheetahs remaining in Iran*

The Evolution of Fish

Rewind to 540 million years ago, to a time called the Cambrian Period. Unlike today, the vegetation then was not so evolved or diverse. In fact, the land was inhospitable for life. However, the oceans were bursting with life and this period was called the 'Cambrian Explosion'. The fossil for the first known vertebrate evolved during this period and was called *Haikouichthys ercaicunensis*. The fossil for the earliest primitive **chordate**, called *Pikaia gracilens*, also dates to this period.

Like an Eel

Pikaia gracilens resembled the modern eels. The animal was small, with a tapered, streamlined body. It had eyes, a brain, and a **notochord**. This refers to a flexible, rod-like structure found in the embryos of all vertebrates. It goes on to become a part of the backbone.

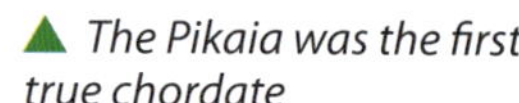

▲ *The Pikaia was the first true chordate*

Jawless Wonder

You might not recognise the first fish to swim on Earth as they had no jaws or teeth. They began to appear in the Ordovician Period, about 488 million years ago. During this period, Earth looked very different from how it looks today. The ocean covered the area to the north of the tropics and land was limited to a supercontinent called **Gondwana**.

The (now extinct) jawless fish of this period were called ostracoderms. Not more than 30 cm in length, they lived in freshwater and had bony scales which protected the brain. They were bottom feeders, which means they found food on the ocean floor.

▲ *The extinct lampreys were jawless fish*

The Jaws Emerged

The *Acanthodians* (or spiny shark) were among the first fish with jaws. They emerged in the Silurian Period, about 440 million years ago. During this period, there were large-scale geographical changes due to the melting of giant glaciers. This resulted in a rise in sea levels.

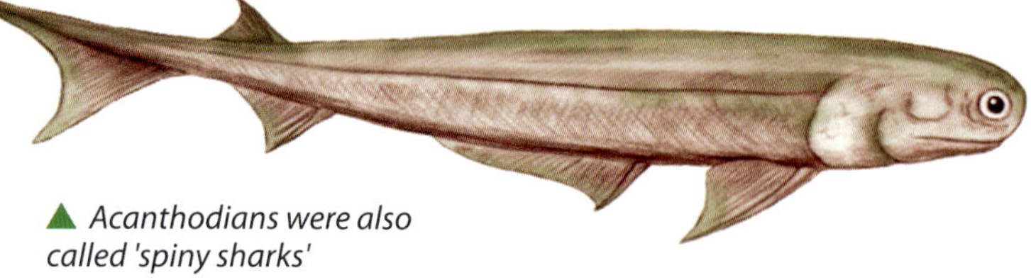

▲ *Acanthodians were also called 'spiny sharks'*

The fish that existed during the Silurian Period survived for about 150–180 million years before they went extinct. Nothing is known about their ancestors, but it is assumed that they originated from the jawless invertebrates.

The *Acanthodians* were small fish with huge eyes and short snouts, indicating that they relied more upon their vision rather than smell. They had bony spines in front of the fins and small, diamond-shaped scales. They had about four fins that they used for manoeuvring rather than locomotion. The latter was managed with the tail.

In Real Life

Sharks are not bony, but **cartilaginous**. Hence, unlike reptiles, mammals, and birds, the skeletons of sharks are made of cartilage and tissue. Human beings also have cartilage in their ears and nose. You might be able to feel it by placing a finger on the tip of your nose. Cartilage has lesser density than bone, and it is more flexible.

The Age of the Fish

The period from 420–358 million years ago is called the Devonian Period. It came to be called the 'Age of Fish' because of the variety and wealth of fish that swam in the waters of this period. Among the many species of fish, the placoderms dominated the waters for almost 70 million years, disappearing at the end of the Devonian Period.

The word 'placoderm' means 'plate skin' and as the name suggests, the fish were covered with bony armour, especially around the head and the trunk. The tail was well-developed. The other important features of this fish type were the presence of jaws, with attached tooth-like structures; and the development of well-formed fins at the pectoral and maybe even the pelvic region.

▶ *Placoderm were generally small fish, but some could be about 13 feet in length. The Dinichthys is an example of a large placoderm species with a size of about 30 feet*

Isn't It Amazing!

Megalodon, meaning 'big tooth', was a huge 25,000 kilogram shark that grew almost 55 feet in length. It had a gaping mouth of 6 feet. Megalodon had sharp teeth which could eat anything, big or small, in the ocean. Each tooth was about 17.7 centimetres in length.

This shark appeared 16 million years ago. It ruled the ocean until two million years ago, when it disappeared. It is now extinct because of extensive changes in the ocean temperatures and lack of prey. But folklore says it still lurks deep in the oceans!

▼ *The megalodon is said to be the largest fish that ever swam the waters of our planet*

▼ *The prehistoric Stethacanthus shark had a strange anvil-shaped dorsal fin on its back. No one knows what it was used for*

Ancestors of Modern Fish

The earliest shark-like fish appeared in the Devonian Period. They were the first cartilaginous fish to appear, belonging to the group called *Chondrichthyes*. While most of these primitive fish disappeared, few still survived and evolved.

The modern sharks, skates, and ray-fish are cartilaginous fish that originated in the Jurassic Period, about 200 million years ago. During this period, the dry, hot climate changed to subtropical and humid.

The cartilaginous fish had flexible jaws. They could battle with fish that were bigger than themselves. A tail fin helped them swim faster and longer so they could easily pursue their prey.

▶ *Helicoprion was another ancient shark. It had a spiral-shaped tooth structure called the tooth-whorl*

Anatomy of a Fish

With over 35,000 fish species in the world, scientists have identified important body parts of the animal. These can help distinguish between the various species that swim in the waters of our planet.

Body

Fish have streamlined bodies which help them move smoothly in water. The shape of the body and the tail have evolved in accordance with their habitat. The shape of the body decides a fish's ability to swim. For example, tuna fish are excellent swimmers and have sharp, forked tails that make them agile.

▲ *Most fish have taste buds all over their body*

▶ *Fish survive in water with the aid of these parts*

Fins

Fish use their fins to swim. The position of the fins on the body determines their functionality. Fish move forward with their tail fins, whereas they use the pectoral fins to swim and balance better. The dorsal fins are used for protection and balance. The ventral and anal fins located on the belly are needed for balance and steering.

Mouth

If a fish has a superior (upward pointing) mouth, it will eat food above it. If a fish has an inferior (downward pointing) mouth, it will eat food from the bottom of the seabed.

Gills

Instead of lungs, fish take in oxygen molecules dissolved in the water through their gills. These are located on both sides of their mouths. They are covered by a series of bones called the operculum. Some fish species have spikes on their gills to protect themselves against predators. Water passes over the tiny network of blood vessels in the gills. The fish force the water to move over the gills using their mouth and throat.

▲ *A close-up of the gills of a fish*

Scales

Most fish have an external covering of scales which act like a protective device. They are unique to fish. Scales come in different sizes and colours; they vary among species and at times even sexes. Like human skin, they protect the fish. The scales are covered by a slimy, mucous layer that protects the fish from bacteria in the water.

Lateral Line

Lateral lines are located under the scales of a fish and consist of many tiny openings or holes. These lines run parallel to the fish's body and allow it to feel low vibrations in the water.

Nares

Nares are located on the snouts. They are the two holes that help a fish smell under water. Some fish have a heightened sense of smell because of the nares.

▲ *The goldfish have nares which allow them to smell better than human beings*

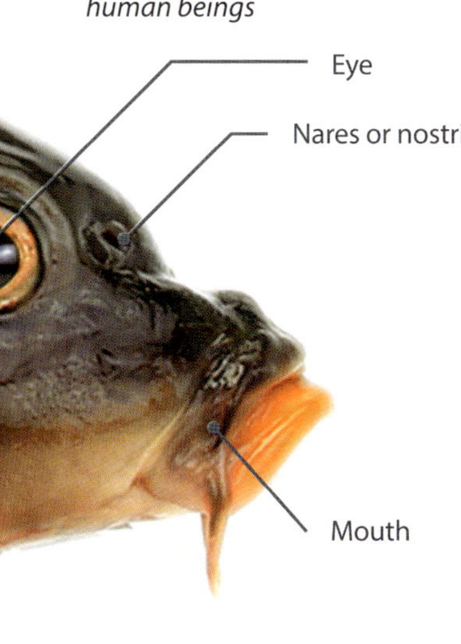

Heart

Fish's internal organs such as the stomach, gall bladder, liver, and kidney are similar to mammals. But their hearts are different. While human hearts have four chambers, fish hearts have only two—one chamber receives blood and the other chamber pumps it out.

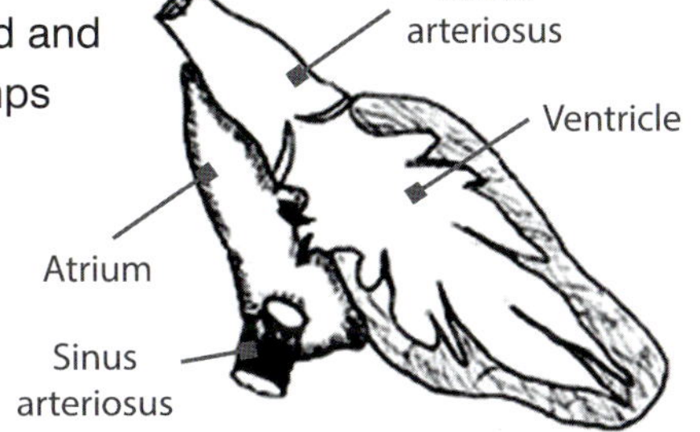

▶ *If a fish's blood pressure is too high, it might damage the gills of the fish*

Gas Chamber

Most fish do not breathe with their lungs. Instead they have a swim bladder, also called the air bladder. This is a gas-filled sac in their abdominal region. It contains a mixture of oxygen, carbon dioxide, and nitrogen absorbed from the blood. It functions as a **ballast** organ, which means it helps the fish maintain depth without sinking or floating upward. The swim bladder amplifies sound in some fish, like the glass fish. This part is missing in cartilaginous fish, deep-sea dwelling fish, and in a few bony fish.

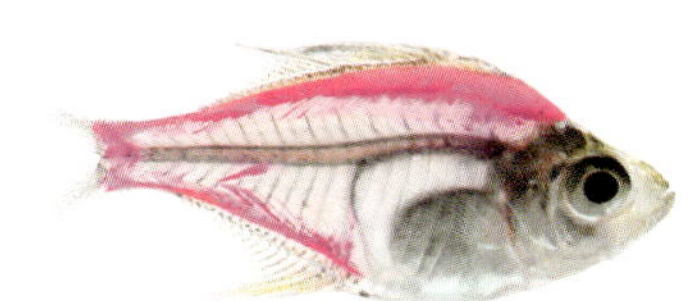

▲ *One can clearly see the swim bladder of a glassfish as it is transparent*

Fishy Features

The word 'fish' comes from the Old English word 'fisc'. It was used for any animal living in water. To us human beings, from where we stand at the highest position on the evolutionary scale, fish might seem like just another set of animals that cohabit Earth with us. But these forms, which are at the base of the evolutionary scale, have many interesting tales to tell.

Cold-blooded

Fish are **ectotherms** or cold-blooded animals. No, their blood is not cold, as the name suggests. It just means that they cannot regulate their body temperatures internally and rely on external sources such as sunlight or shade to maintain it. Most often, their temperatures tend to be slightly lower than that of the environment to prevent loss of moisture.

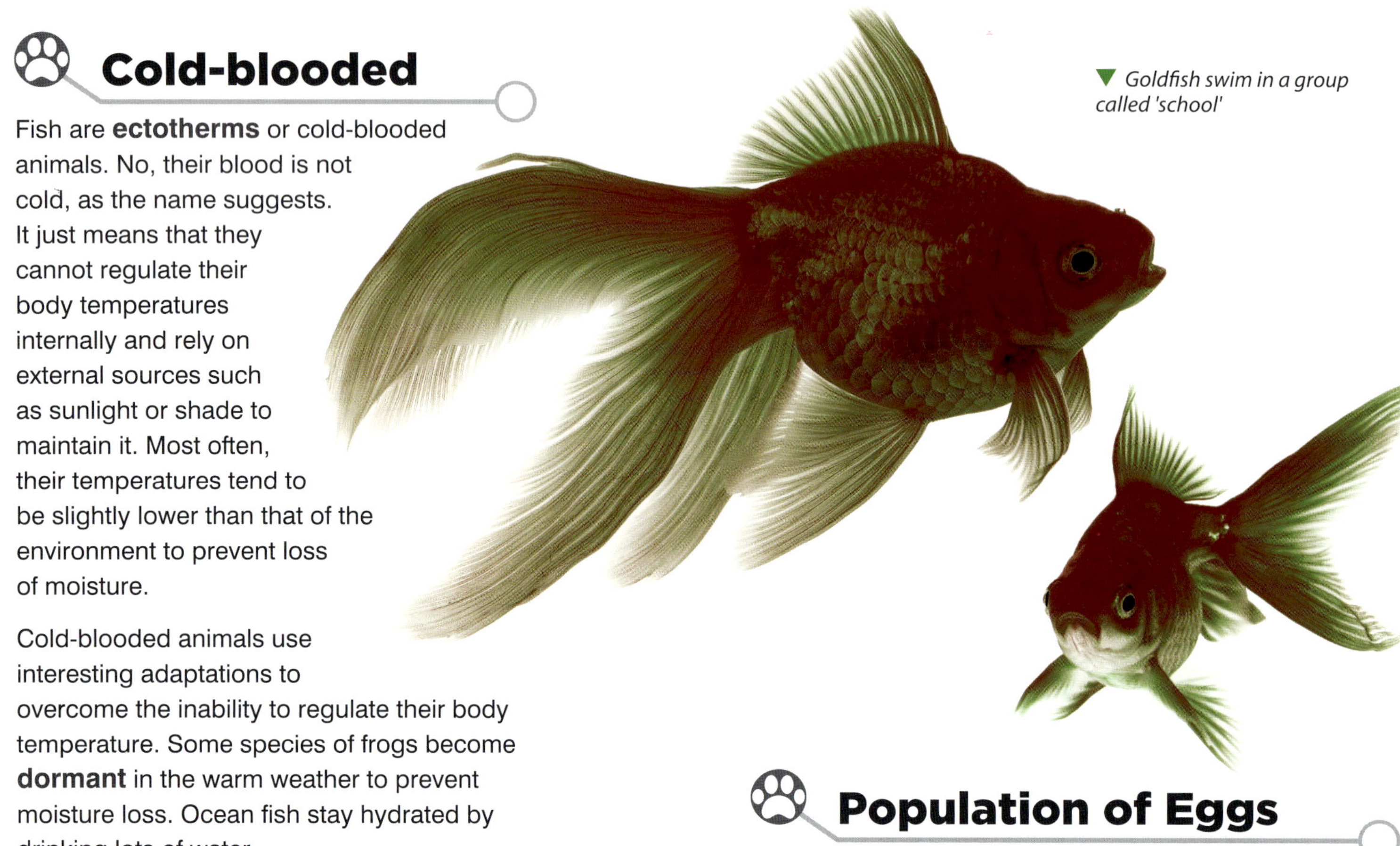

Goldfish swim in a group called 'school'

Cold-blooded animals use interesting adaptations to overcome the inability to regulate their body temperature. Some species of frogs become **dormant** in the warm weather to prevent moisture loss. Ocean fish stay hydrated by drinking lots of water.

Vertebrates

It is with the fish that the world got its first vertebrates. A vertebrae or backbone supports the body and protects the spinal cord. So, vertebrates need their backbones and brains to survive.

Laying Eggs

Fish lay their eggs in water. However, unlike mammals, birds, and reptiles, they do not lay amniotic eggs, that is, eggs containing **amnion**. Amnion is a fluid-filled sac that keeps the embryos moist.

A siamese female fighting fish guarding her newly laid eggs amongst the bubble nest

Population of Eggs

Fish lay several eggs at one time. This ensures that the reproduction process is successful. In fish, fertilisation takes place outside the body. In most cases, the fertilised eggs are not looked after by the adults. This exposes the eggs to predators and other dangers that could reduce the number of eggs drastically. So, not all eggs lead to successful reproduction. Those that fail to hatch are recycled into the ecosystem.

Isn't It Amazing!

A unique feature of the clownfish is that they are all born male during birth. If their group is missing a dominant female, one clownfish will change its sex. They cannot change back to male. This is because a school of clownfish follows a hierarchy. The most dominant female rules the school. So, if there is no dominant female, the dominant male changes sex. Similarly, moray eels and gobies also change sex.

Bony v/s Cartilaginous Fish

A distinct feature of the fish evolution cycle was the early development of bone, cartilage, and enamel-like substance. This helped the fish adapt to diverse aquatic environments in the water and even on land.

Differences

There are many differences between the bony fish and the cartilaginous fish. The most important difference is that the skeleton of the bony fish is made of bones, whereas the skeleton of cartilaginous fish is made of cartilage.

Features	Bony Fish	Cartilaginous Fish
Skeleton	The **endoskeleton** is made of bones. Their **exoskeleton** or external covering is made of cycloids (thin, bony plates).	The endoskeleton is made of cartilage. Their exoskeleton is made of placoid, that are small denticles with a sharp, enamel coating.
Mouth	They have an anterior tip mouth opening.	They have a ventral mouth.
Operculum	They have opercula on either side of the gills.	They do not have an operculum.
Tail	They have a homocercal tail fin.	They have a heterocercal tail fin.
Swim bladder	Also called an air bladder or buoyancy organ, it is present in most bony fish. It contains oxygen and allows the fish to maintain their depth without floating upwards or sinking.	It is absent in all cartilaginous fish and bottom-dwelling fish.
Habitat	They are found in both freshwater and marine water.	They are predominantly present in marine waters.
Reproduction	They externally fertilise their eggs.	The fertilisation is internal.
Excretion	They excrete ammonia.	They excrete urea.

Flying Fish

The flying fish is a bony fish commonly found in warm seas. They consume a variety of small creatures, but mainly plankton and measure 17–30 centimetres in length. These fish have pectoral fins that are like a bird's wings. They use these fins to easily glide over water. They usually swim in schools and consist of over 40 different types of oceanic fish. Some species such as the blue flying fish have two wings, whereas some are four-winged, like the California flying fish. They are preyed on by dolphins, tuna, and mackerel. They avoid these predators with their aerial skills.

▲ *Illustration of the flying fish*

Flying fish do not generally fly, but rather glide on their fins at the speed of about 59 kmph. They can easily cover a distance of 200 metres. The flying fish build up speed under water, then become airborne, making many consecutive glides with their tails propelling their movement above water. The strong fliers can even go as fast as around 400 metres in one single glide!

◀ *On an average, flying fish live up to 5 years*

Dogfish

Dogfish is an example of a cartilaginous fish. It travels in a dense school of fish and preys on other fish and invertebrates. The spiny shark has a sharp spine in front of each of its two dorsal fins. The dorsal fins have venom glands that can cause painful injuries. It is between 60–140 centimetres long. People kill dogfish for food or to produce shark liver oil. They have also been used to make fertilisers.

▲ *Dogfish have been considered a nuisance because they end up ripping fishnets after taking bait*

Predatory Sharks

Sharks—for some, hearing the word itself sends chills down the spine. In reality, most sharks do not attack humans unless they feel threatened. However, human beings hunt sharks for their fins, teeth, meat, and skin. So human beings do pose a threat to their survival. There are over 500 species of sharks in the world today.

Behaviour and Intelligence

Although considered to be instinct-driven hunters, many species of sharks have exhibited complex problem-solving skills and high curiosity. They have also shown signs of playful behaviour by repeatedly rolling around in kelp and chasing a trail in the ocean for fun. If a shark feels threatened, it exhibits exaggerated swimming movements to ward off predators. The intensity of the movements is directly dependent on how threatened the shark feels.

Sleep Patterns

Some sharks sleep like dolphins, with one part of their brain active and awake, while the rest of the brain rests. Some sharks use their spiracles to continue swimming while they sleep. Their spinal cord coordinates the activity of swimming so that the brain can rest. Some sink to the bottom of the ocean and lie still with their eyes open and actively pump water over their gills.

▲ *A scuba diver is seen swimming with the whale shark. These sharks are harmless, but they can be spooked by too much movement or artificial light*

▶ *There are 10 different species of hammerhead sharks. They vary in size but have the same basic features*

Hammerhead Sharks

Hammerhead sharks are cartilaginous animals. Each hammerhead shark is 13–20 feet in length and weighs anywhere between 230 and 450 kilograms. What makes this shark unique is the shape of its head, which looks like a flattened hammer. With this peculiar shape, the hammerhead shark can lift and turn its head quickly.

Their wide-set eyes give them a wider view of the surroundings. The hammerhead sharks have expanded nostrils, thereby giving them a better sense of smell and better ability to locate the prey than other fish. Moreover, the underside of the wide head has **electroreceptive organs**, which help these sharks detect the electrical impulses given off by their prey. Great hammerhead sharks are greyish brown or olive green on top, with a white underside.

▲ *The scalloped hammerhead shark has been listed as Critically Endangered due to overfishing for its fins*

Hunting

The great hammerhead is the largest of the 10 identified species of this shark. It has sharp, blade-like teeth. It preys on larger fish, squids, crabs, smaller sharks, and its favourite, the stingrays. As a matter of fact, the shark uses its electroreceptive organs to hunt for stingrays, which usually bury themselves under sand. They use their enormous heads to hold the stingrays down during a fight.

Habitat

The sharks are present in tropic or temperate marine waters. They feed around shallow coastal areas, estuaries, and salty waters. In winters, huge schools of hammerhead sharks are seen migrating towards the warmth of the equator and in summer they move towards the poles.

Young Ones

All species of hammerhead sharks are viviparous, which means that the females retain fertilised eggs within their bodies to give birth to live young ones. The female great hammerhead shark gives birth to several dozens of young ones at a time. The females give birth in shallow waters in summer and spring. The young ones continue to stay there until they are old enough to venture into the oceans.

▶ *Sharks have been around for more than 400 million years*

A Poisonous Defence

Whether they live in the ocean or a large aquarium, fish make the world beautiful. They are also great at defending themselves against predators. Quite a few of these creatures have defence mechanisms that could harm not just their predators, but also human beings.

Sting like a Stonefish

A stonefish is about 30 centimetres in length, with a large head and mouth, and small eyes. These details sound ordinary, but the fish is extraordinary. It carries venom in its skin and in sacs attached to the razor-sharp spines along its back.

When attacked or accidentally stepped onto, the stonefish pushes the spines into the predator and releases the poison. The sting of the spine can lead to paralysis or even death in a few cases. These fish are bottom-dwellers; they live in estuaries, rocks, or in the corals found in the world's oceans.

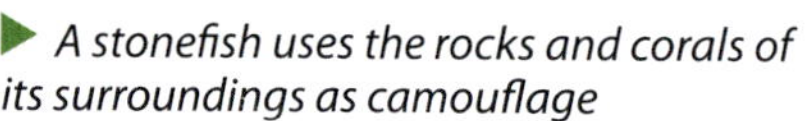

A stonefish uses the rocks and corals of its surroundings as camouflage

All Puffed Up

Pufferfish, also known as blowfish, have the amazing ability to expand. They swim clumsily, making them an easy target for predators. But to escape, the fish fill their elastic stomach with water, and at times air, to become an inedible ball several times larger than their size. Some species of pufferfish have spines on them, making them even less palatable.

The next line of defence the pufferfish uses is tetrodotoxin, a deadly toxin present in almost all 120 species of pufferfish. It is so lethal that the predators, including humans, barely stand a chance for survival. There is enough poison in a single pufferfish to kill almost 30 adult human beings. There is no **antidote** available to fight the toxin.

The only species immune to pufferfish toxin are sharks

Fatal Red

The average length of a red lionfish is about one foot. It lives amongst the rocks and crevices of the Indian Ocean and the Pacific Ocean, but it has been introduced to warm waters worldwide. The red lionfish looks striking with brown, red, and white stripes.

It has around 18 venom-filled, needle-shaped dorsal fins. If disturbed, it spreads its fins and if further intimidated, it attacks. If a human being is attacked by the sting of the red lionfish, they might feel nauseous and would not be able to tolerate the pain, but it is rarely fatal.

The red lionfish is a prized aquarium fish

Spectacular Tentacles

Octopuses are molluscs. They are called cephalopods, which means 'head-footed'. Unlike snails, in case of octopuses, their molluscan foot is located at the headend. The animal is named octopus because of the eight distinct arms of the invertebrate.

Facts of Physique

Octopuses have soft, round, sac-like bodies. They have large bulging eyes and a beak-like mouth hidden in the ring of arms. The powerful arms are covered with suction cups called suckers. These are used to catch prey, move over rocks, and swim.

An interesting fact about this invertebrate is that it has three hearts and nine brains. Two of its hearts supply blood to the gills, the breathing organ. The third heart is the organ heart, which supplies blood to the other organs.

▲ *Female octopus have to keep their eggs safe from predators as well as gently guide currents of water over them, so that they get a fresh supply of oxygen. They do this without leaving their side or eating. When the eggs hatch, the mother is so exhausted and starved, that she dies*

▲ *On mating, the female octopuses lay eggs on rocks. They are small, shaped like a teardrop and milky in appearance*

Blue-blooded

The blood of an octopus is blue in colour because of the presence of hemocyanin, a copper-rich protein. This protein binds to the oxygen molecules, thus helping transport oxygen throughout the body.

Intelligent Beings

Octopus brains are another story. One of them, the big one, controls the **nervous system**, while the remaining small brains control the animal's tentacles. Long ago, octopuses were thought to be dumb, but did you know they are one of the most intelligent invertebrates? They are known to recognise different colours, shapes, carry out simple tasks such as opening boxes, navigate through mazes, and solve simple problems.

Dwelling and Feeding

Octopuses are solitary creatures. They build dens with rocks that they move with their arms. Few do live near the water's surface, but most octopuses are deep-sea dwellers. They crawl on the ocean floors using their arms, looking for food to eat. They rise at dawn and dusk to look for food along the water's surface. Their favourite meals include crabs, shrimps, and lobsters.

Camouflage

Octopuses have pigment cells and special muscles which help them not just change colours but also match the texture and pattern of their surroundings. Sharks, dolphins and many other predators simply swim past such a camouflaged octopus.

Defence

If a camouflaged octopus is discovered, it squirts an ink-like liquid. This darkens the surrounding water, dulling the predator's sense of smell, giving the octopus enough time to escape. If this fails too and the octopus is caught by the predator by its arm, it just lets go the body part. Yes, it can regrow its arm! The beak-like mouth can also give a nasty bite to the predator and the saliva too is filled with venom.

Evolution of Amphibians

The fossil record of amphibians has been poor until recently. However, new discoveries give us a clear picture of the history of amphibians. As the Silurian Period went on, the diverse species of fish continued to evolve. Scientists do not know for certain if they got tired of fighting off too many predators or if they had too much competition for food, but eventually some fish began to move towards land.

▲ *An extinct relative of the modern scorpion, called Slimonia, was one of the earliest animals to live on land along with mites and spiders*

The Missing Pieces

The first animals to move to land and settle there were the arthropods. They had started their journey towards land 100 million years ago. The bodies of these animals were uniquely adapted to life on land as they had strong legs, light bodies, and hard **exoskeletons** that helped them conserve water.

These animals had to figure out how to conserve water, move, stay safe from predators, exchange gases (like oxygen and carbon dioxide), and learn to support themselves against the strong pull of gravity.

The Story Begins

During the Devonian Period, land was occupied by arthropods and **tetrapods**. It is said the amphibians evolved from tetrapods and were related to the primitive **lobe-finned fish**. These fish probably had lungs that helped them breathe on land. Their bony limbs had digits at the ends that helped them crawl on land.

Eusthenopteron is a fossil belonging to the late Devonian Period. Scientists believe this animal was near the main line of evolution from fish to amphibians. It was about six feet long, had a broad skull with teeth, and was a carnivore. The vertebral column was not well developed, but the fins had bony structures for support.

▲ *A beautifully preserved fossil of Eusthenopteron*

Eusthenopteron was not built for life on land. It probably lived in shallow waters, climbing small rocks and plants in search of food.

Isn't It Amazing!

In the Late Triassic Period, amphibians reached gigantic proportions. There lived a giant salamander called *Metoposaurus algarvensis*. It was the size of a small car and perhaps even ate small dinosaurs!

Paving the Path

The fossil of an aquatic animal called *Tiktaalik roseae* was discovered in 2004 in Canada. It was a tetrapod and about nine feet long. It had sharp teeth and, unlike fish, it could move its head from side to side to look for prey and predators. It probably had arms with elbows, wrists, and shoulders. Hence, it is considered to be the first fish with limbs. It had a flat skull with bulging eyes like that of a modern crocodile, suggesting it swam beneath waters in lakes, streams, and swamps.

▶ *An illustration of the Tiktaalik roseae*

▶ *Ichthyostega were more advanced than Eusthenopteron and closer in relation to the first tetrapods*

Moving Towards Land

The fossil of *Ichthyostega*, an animal closely related to the tetrapods, was found in Greenland. It belonged to the late Devonian Period. The animal was about three feet in length. A unique feature of this animal was that each of its feet had seven toes. It did show aquatic traits such as a short snout, presence of a bone in the cheek region that covers the gills in fish, and small scales on the body. But, like tetrapods, it lacked gills, had bones supporting fleshy limbs, and strong ribs. *Ichthyostega* could leap and move on land.

▲ *Being short with broad limbs might have helped the Eryops walk on land*

Landed

About 340 million years ago, the evolution of the first known family of amphibians took place. They were called the *temnospondyls*. *Eryops* is an example of this group. It had a stout body, large skull, and a strong vertebral column, suggesting the animal could move on land. Its teeth were sharp, indicating that it could have been a carnivore.

What Happened to Ancient Amphibians?

Ancient amphibians were one of the largest predators on land for millions of years. But in the Permian Period, about 280 million years ago, Earth's climate changed. This period marks the largest mass extinction in Earth's history. Wetlands and swamps were replaced by vast deserts. But amphibians needed water to reproduce and survive. The change in climate caused many of the amphibian species to die out.

Extinction and Resurgence

The Permian Mass Extinction was an event which wiped out more than 95 per cent of all life on Earth. This included the giant amphibian predators. However, a few survived in the shadows of the reptiles (dinosaurs).

During the Jurassic Period, the climate changed once again, giving rise to wetlands, and the surviving amphibians evolved to form the modern amphibians we know today, such as frogs, salamanders, and newts.

Incredible Individuals

Jodi Rowley is a biologist who studies amphibians, focusing on their diversity, environment and the threats to their existence. Today, it is said that one-third of all amphibian population is threatened with extinction and Jodi Rowley is taking measures for their conservation. She is mainly focussed on the conservation of Southeast Asian amphibians.

Characteristics of Amphibians

To us human beings, from where we stand at the highest position on the evolutionary scale, amphibians seem like just another set of animals that cohabit Earth with us. But these animals, which are at the base of the evolutionary scale, have many interesting tales to tell.

The Heart of the Matter

Fish and amphibians have similar structures for many internal organs such as the stomach, gall bladder, liver, and kidney. But their hearts are structured differently. A fish has a two-chambered heart, where one chamber receives blood and the other chamber pumps blood out.

On the other hand, amphibians have a three-chambered heart. The first chamber receives the oxygenated blood, the second chamber pumps it to the entire body, and the third chamber receives deoxygenated blood.

Amphibians hearts are like balloons where they can only take up a limited amount of gas. So, they diffuse the extra oxygen through their skin.

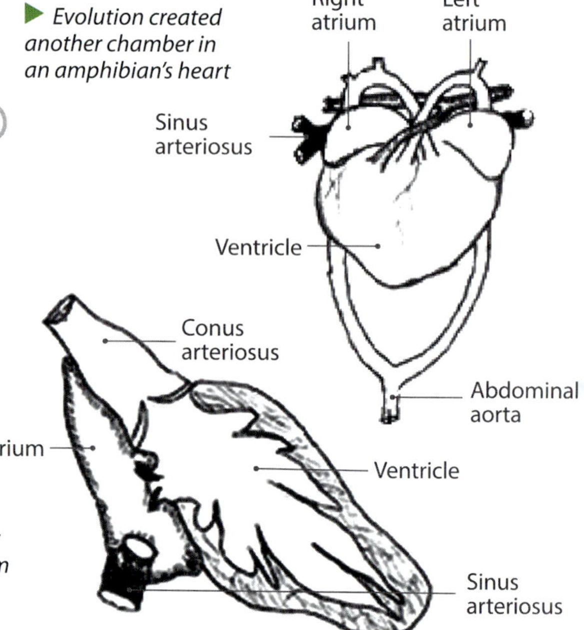

Evolution created another chamber in an amphibian's heart

A fish's heart has only two chambers: an atrium and a ventricle

Spitting Frogs

Frogs can swallow insects whole by simply pushing their sticky tongues out. The saliva helps them capture the prey. To swallow the insect, a frog closes its eyes. This helps it push the insect down towards the throat. However, when a frog eats something poisonous, it simply spits out the contents of its stomach.

Yes, you read that right, the food in a frog's stomach can come out entirely through its mouth to discard the disagreeable contents. When a frog eats bombardier beetles, the insect might squirt boiling chemicals into the frog's stomach as a defence, so the frog will spit it out.

The saliva on the long, sticky tongue of a frog acts as glue for the insects

The Blind Dragon

The olm is a blind salamander that lives in caves. It resorts to its other sensory organs to hunt for food. It has pale pink skin. It does not grow, instead it maintains its red feathery gills that form when it is a **larva**, even after it reaches maturity. It was once called a baby dragon because of the small size of its snake-like body. Olm salamanders can live for more than a hundred years.

The olm salamanders have existed for nearly 20 million years in the caves of Croatia and Slovenia, but chemical pollutants are now threatening the ancient animal's existence.

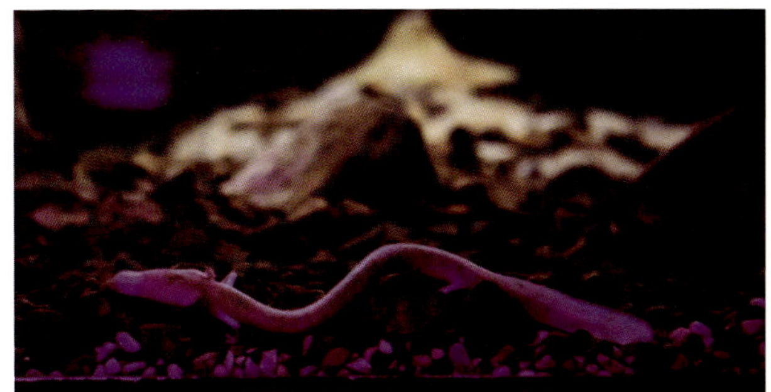

▲ Olm salamanders eat crabs, snails and small insects

Most Moist

Amphibians have a moist, slimy skin as seen in most frogs. This is one of the reasons amphibians stay near water. Few amphibians live in hostile environs, such as the African bullfrogs who endure the hot, harsh climate of southern Africa. They have skins with long ridges to prevent moisture loss.

▲ Have you ever seen a frog look dry? They always have slimy and moist skin as they need to stay moist for survival

Commonly seen in amphibians is the process of metamorphosis. As the immature larva develops, it undergoes bodily changes that make it distinct from the adult it grows into. For example, most amphibians spend the early years of their lives in water, but as adults they take to land.

◀ A close-up of the larva that frogs develop from

▲ Once the insect is in their mouth, frogs use their eyeballs to swallow it. The eyeballs move to the mouth cavity and push down the food, exerting force that liquifies the saliva and releases the insect

To Catch a Fly

Research shows that frogs might have special spit that helps them catch flies. Their saliva is sticky, so frogs can extend their tongues and hold onto the flies that they catch. Their saliva, which is normally thick, liquefies to spread over the insect's body and then thickens again, effectively trapping it.

Incredible Individuals

In the 17th century, a woman named Catharina Geisslerin in Germany claimed that she had swallowed tadpoles from a swamp. She claimed that these tadpoles were happily living in her body as frogs. She also claimed that drinking milk caused these frogs to jump out of her mouth. Surprisingly, she managed to duplicitously convince physicians that this was true by vomiting frogs in front of them.

Evolution of Reptiles

The reptiles that we see on Earth today can be traced back to small lizards. These lizards roamed the planet more than 350 million years ago. They laid amniotic eggs which allowed them to move from water to land. Slowly, these land vertebrates grew in size.

320 Million Years Ago

The reptiles separated into two groups—**synapsids** and **sauropsids**. Synapsids were the early ancestors of modern mammals. Sauropsids evolved into modern birds, reptiles and dinosaurs. Though both were vertebrates, for a few million years, the synapsids became the dominant species. *Hylonomus* and *Paleothyris* were the earliest reptiles.

▲ *Hylonomus were small, preferred to live in swamps and fed on insects*

245 Million Years Ago

The Permian Mass Extinction occurred over 252 million years ago and lasted for nearly 15 million years. It wiped out 95 per cent of marine animals and 70 per cent of land animals, including most synapsids. However, the sauropsids survived and even thrived while the synapsids declined in number. This came to be known as the **Triassic takeover**.

The Mesozoic Era

Large dinosaurs gradually evolved from sauropsids nearly 225 million years ago. They dominated the land and became one of the most important groups of animals of the **Mesozoic Era**. This era was divided into the Triassic Period, the Jurassic Period, and the Cretaceous Period.

During the Triassic Period, both dinosaurs and reptiles evolved rapidly, but it is only during the Jurassic Period that the diversity of the dinosaur species evolved. Many dinosaurs became extinct towards the end of the Mesozoic Era, but the reptiles survived.

230 Million Years Ago

Evolving from their small lizard-like ancestors, reptiles began to rule not only on land but also in air. The major predators of this time were the Archosaurs. They were ancestors of the present-day crocodiles but were very different in appearance. Most animals belonging to this species had strong forelimbs and long hind legs. They had two skull openings, one in the snout and another in the lower jaw. But the function of these openings is unknown.

▼ *Proterosuchus was the first template for the modern crocodile. It had a slight downward curve in the upper jaw*

▶ *Archosaur means the 'ruling reptile'. They were divided into the Pseudosuchia (crocodiles) and the Ornithosuchia (birds)*

The Flying Reptiles

Pterosaurs, or 'flying reptiles', ruled the skies for about 150 million years. They should not be confused with dinosaurs, but they did belong to the group of archosaurs. They had no competition till the birds arrived millions of years later, as they were the first flying vertebrates.

Pterosaurs had light, hollow bones and their wing surface was formed by a membrane of skin similar to that of the modern bats. This membrane stretched across their thin fingers. Another membrane stretched from the wrist to their shoulder. This allowed them to fly with ease. They had small, sharp needle-like teeth in their jaws with which they could catch fish while still in flight.

▲ Archelons were about 12 feet in length

▲ Some scientists believe pterosaurs walked on their hind legs, while others believe they walked on all fours with their wings folded by their sides

Dominating the Seas

In the oceans, marine reptiles had become the dominant species. The long-necked Elasmosaurus was one of the largest sea creatures that had needle-like teeth perfectly suited for catching fast-moving fish. The first turtles evolved during the Cretaceous Period. The Late Cretaceous Period had turtles called Archelons. They protected themselves from predators with shells that were similar to those of the modern sea turtles. They moved in the water by pushing themselves forward with their front feet. Archelon died out with the dinosaurs around 65 million years ago.

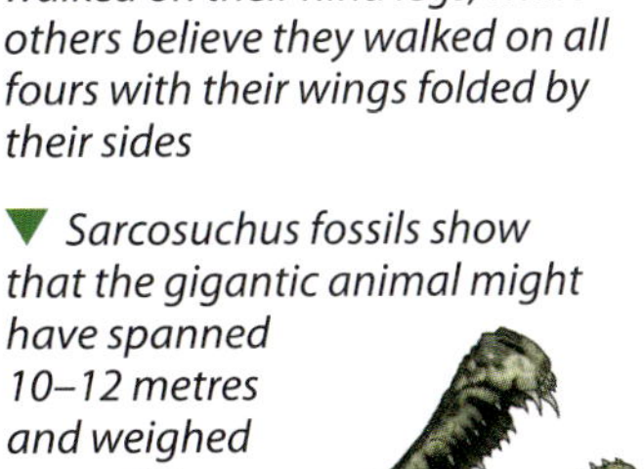

▼ Sarcosuchus fossils show that the gigantic animal might have spanned 10–12 metres and weighed around 8 metric tons

Evolution Continues

Crocodiles and lizards continued to survive and evolve alongside dinosaurs. The crocodiles grew to extremely large sizes, such as the Sarcosuchus which preyed on some species of dinosaurs. These monstrous crocodiles died out with the dinosaurs.

Some lizards lost their legs and became the first snakes. These snakes were mostly constrictors, like the modern pythons, but smaller. At some point during the evolution process, the 'Sonic hedgehog' gene switched off in snakes. This gene is necessary for the growth of limbs, and gets its name from the spines that grow from the embryo during the limb-development stage. So, snakes became limbless reptiles. However, a genetic mutation could cause them to grow limbs once again.

◀ While most were small, the Titanoboa is the largest known member of the serpent family. They lived near the end of the Cretaceous Period, and weighed around 1,135 kilograms, and had a length of roughly 13 metres!

Reptilian Features

The word 'reptile' refers to the animals' creeping and crawling movement. Reptiles are vertebrates, which means they have backbones which support and give shape to their bodies. They live in hot deserts, swamps, oceans, and forests. Animals can only be classified as reptiles if they possess certain characteristic features.

Cold-blooded

Reptiles are ectothermic or cold-blooded animals. Like amphibians and fish, reptiles cannot regulate their body temperature. They need sunlight and shade too. That is why, in their natural habitats, chameleons have been observed basking on hillside rocks in winters. Since they are cold-blooded, reptiles cannot survive without heat.

They lie perpendicular to the Sun, so that they can absorb a lot of sunlight. If it gets too hot, the reptile will lie parallel to the Sun or look for a shady spot. A chameleon will lighten the colour of its skin if it gets too hot, a crocodile might keep its mouth wide open, while a snake might burrow into the soil.

Reptiles in the desert region, such as Gila monsters, are nocturnal in summers because their bodies cannot handle the harsh heat during the day. However, if the temperature is tolerable, they can be diurnal.

Tough Skin

Having tough, scaly skin is the key to survival in reptiles. The skin not only acts as a protective covering, but also prevents loss of water, which saves it from drying up. The scales are made of keratin, the structural protein present in the hair and fingernails of human beings. The scales may be tiny—as found in dwarf geckos, or large—as seen in lizards. Even turtles have scales called scutes, which are arranged in a staggered manner. The scales on a crocodile are called armour.

Incredible Individuals

Back in 1992, Steve Irwin filmed a documentary with his wife, Terri Raines, called *The Crocodile Hunter*. It became so popular that he was given the opportunity to film more of the documentary, of which currently there are 50 episodes. 'Crocodile Hunter' became his nickname, even though he never actually hunted a crocodile. In fact, he was known for his close encounters with dangerous animals and his work with conservationists to help save the crocodiles.

A turtle's shell has scutes made up of keratin. Though they act as the shell, they are built like a nail, beak, or horn

Eggs

Most reptiles lay eggs. Without the special feature of the amniotic eggs, reptiles could never have moved onto land. These types of eggs have leathery or hard shells which protect the embryo. The eggshell allows the flow of air but prevents vital fluids from leaking out. Inside the eggs, there are a series of fluid-filled membranes which help the embryo survive and develop.

▲ *Like most creatures that are born in eggs, turtles also have egg teeth. They help the baby creature to break out of its shell*

▼ *Crocodilians and birds are the only surviving dinosaurians*

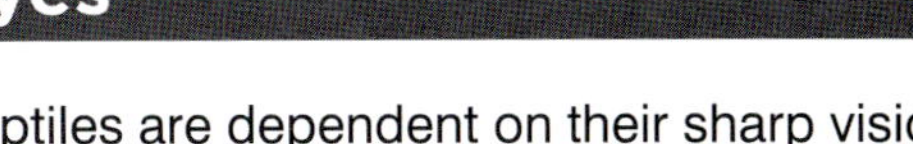

Eyes

Reptiles are dependent on their sharp vision to catch prey. Crocodiles have eyes specially adapted for hunting at night, while chameleons can move their eyes independent of each other. This helps them scan their surroundings for extremely tiny insects. However, snakes such as pythons and rattlesnakes have poor eyesight. They use what is called 'heat vision', which helps snakes sense the heat given off by warm-blooded creatures, through small openings in front of their eyes.

▲ *Chameleons sit patiently, waiting for their prey to come close. As soon as they see their prey, they stick out their tongues and grab it*

Teeth

Reptiles have no specialised teeth such as canines, incisors or molars, like mammals do. They have a long row of conical teeth to catch and chew the prey, as seen in crocodiles and alligators. These animals either swallow the prey whole or cut it into bite-size portions before swallowing. Turtles have sharp edges on their jaws, which they use to bite off seaweed and other foods. Snakes have fangs. The number, placement and shape of teeth varies among reptiles.

In Real Life

The incubation temperature among crocodiles and some turtles, at a certain stage of development, determines the sex of the embryo. For example, in red-eared slider turtles, if eggs are incubated below 28° C, the hatchlings will be male. If the temperature is above 31° C, the hatchlings will be female, and in the temperatures between, it could be either male or female.

Activate Defence Mode

Reptiles and amphibians aggressively prey on animals and defend themselves with ease. Many small reptiles fall prey to carnivorous birds, mammals, and even larger reptiles. To protect themselves from danger, small reptiles such as lizards, snakes, and turtles have devised interesting methods of defence. Similarly, amphibians also have special adaptations that keep them safe.

Ball Up!

The armadillo girdled lizard is found in South Africa. This animal has a unique method of defence. The moment it senses danger, it curls its body into the shape of a ball by holding its tail in its mouth. For attackers, such as snakes, it becomes difficult to swallow the armadillo girdled lizard, not just because they do not know how to take the ball into their mouth, but also because it has a hard, bony covering, and a spiny tail and head. The lizard remains in this shape till the danger passes.

An armadillo girdled lizard in the defence pose

The Australian frilled lizards are threatened by birds, dingoes, and snakes

A Frilled Fight

The Australian frilled lizard uses many tactics to scare off its predators, such as opening its mouth wide, which causes the skin flap to open on both sides of its mouth, making it appear like the lizard has a frilled collar. It also lashes its tail back and forth and makes hissing sounds.

The Mimic

Scarlet kingsnakes are a smart variety. As part of a defence mechanism called 'mimicry', these harmless snakes sport bright colours similar to the venomous coral snake. Seeing the colours, predators mistake them for the dangerous coral snakes and stay away.

These snakes are around 30 to 50 centimetres in length

Hognose

Hognose snakes feign death if threatened. They roll on their backsides and hang their mouths open. They also emit smells similar to decaying flesh. The predator believes that the snake has been poisoned or diseased and moves away.

Hognose snakes are found in North and South America

In Real Life

Hognose snakes have mild venom, but they are not constrictors. They swallow their prey (like toads) whole. Their venom is not toxic to human beings. They tend to be relatively calm and do not bite trainers. Hognose snakes are shy, so their first defence is to hide from their predators. They hide in burrows or leaves. They are often taken captive and are easy to care for.

Hinge-back Tortoise

Apart from stiff, bony shells which are the first line of defence in tortoises, few species also have hinges on the shell. When threatened, the hinges allow the front and back of the animal to close tightly, protecting its soft body parts. Not just this, the hinges also give a painful pinch to the predator.

▲ *The hinge-back tortoise likes to look for water underground by burying its head into the land when the water is sparse*

Incredible Individuals

In Malaysia, there was a famous snake charmer named Ali Khan Samsuddin. He was nicknamed 'King of the Snakes'. Often during his exhilarating shows, he was seen kissing snakes like the king cobra. He once lived in a glass cage with 400 snakes for 40 days. When he was 21, he was first bit by a king cobra but survived. Then again at the age of 48, he was bit by another king cobra, leading to his demise.

A Horned Defence

Thorny devils are adapted to the Australian outback heat. They have moisture-attracting grooves between the scales. During nights, when dew condenses on their bodies, the grooves take in the moisture and send it to the mouth. This helps them survive harsh climates.

These lizards have prickly horns protruding from their skin. They protect the lizard from being eaten, as the predator stays away seeing the horns. If this does not work, they squirt out blood from tiny vessels near their eyes to scare off the attacker.

▲ *Thorny devils survive on ants*

Common Snapping Turtle

Common snapping turtles often remain buried in muddied shallow waters. They are not aggressive in water, but on land they move forcefully towards the attacker and try snapping their mouths aggressively.

▲ *Do not be fooled by its appearance, a turtle can have a sharp bite*

Five-lined Skink

The five-lined skink wriggles its tail to distract predators. The vertebrae of their tails have special fracture points. If pulled from these points, the tail falls off. The attacker's attention goes to the still-wriggling tail. It lets the lizard out of its mouth and makes a grab for the tail. The lizard escapes and eventually grows a new tail.

▼ *These lizards have grey, brown or black skin*

HUMAN BODY

The human body is a remarkable and complex biological organism made up of different systems that work together to maintain life. These systems include the nervous system, circulatory system, immune system, respiratory system, muscular, and skeletal system and digestive system.

The nervous system is responsible for coordinating and transmitting signals and information throughout the body. It includes the brain, spinal cord, and network of nerves. The circulatory system is responsible for the transportation of oxygen and nutrients to different parts of the body, while also removing waste products. It comprises the heart, blood vessels, and blood.

The muscular and skeletal systems work together to allow movement and support the body. They form the framework that holds up the body and the muscles that allow us to move. The digestive system is responsible for breaking down and absorbing nutrients from the food we eat and getting rid of waste. It includes the mouth, esophagus, stomach, intestines, and associated structures.

The immune system is not a system but protects the body against foreign invaders, such as viruses, bacteria, and parasites. It consists of white blood cells, lymph nodes, and specialized cells and proteins that work together to defend the body against harmful pathogens. The Respiratory system is responsible for providing oxygen to the body and removing carbon dioxide. It includes the lungs and associated respiratory structures.

All these systems work seamlessly to keep the body functioning and healthy allowing us to perform a multitude of activities. Understanding how they function is an essential part of maintaining a healthy and active lifestyle.

Our brain remakes the universe inside our heads

Command and Control
The Brain, Spine, and Nerves

The nervous system is to our body what the government is to a country. The nervous system takes the information coming from various sensory organs, processes it and makes decisions, which are then carried out by other organs. The nerves, which are a part of the nervous system, are made up of special cells called **neurones** that take tiny electrical signals from the sense organs to the brain and from the brain to the motor organs. Then the motor organs, mainly the muscles, act together to do what the brain wants them to. The neurones communicate through tiny switches called **synapses**, and chemicals that doctors call **neurotransmitters**.

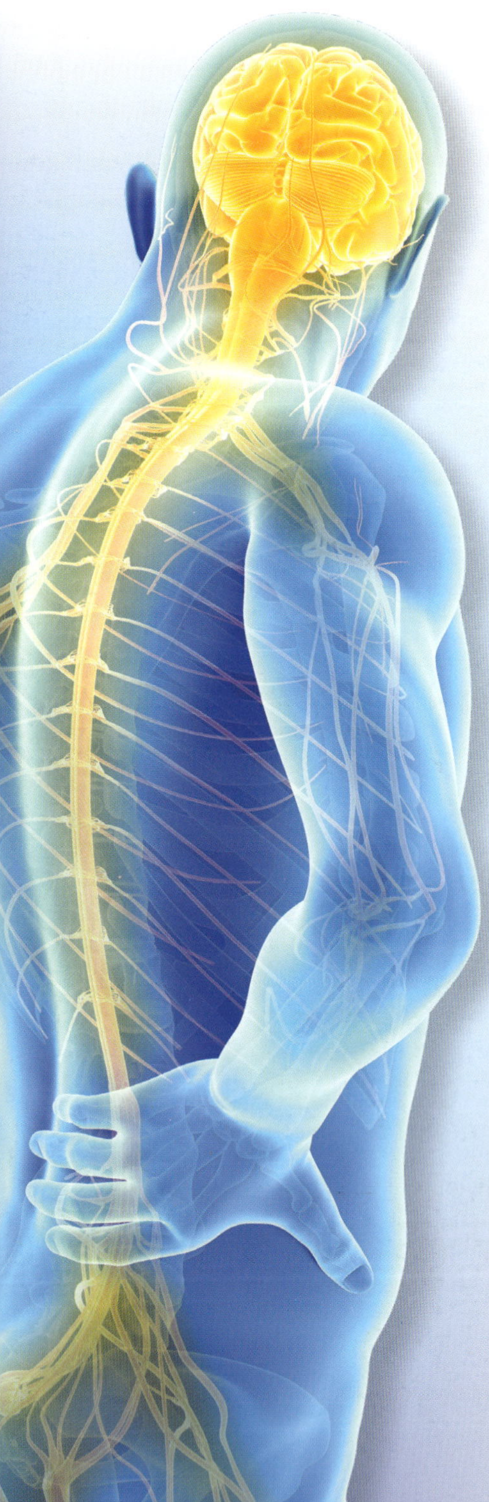

The Brain

The human brain is made of **white matter** and **grey matter**. As human beings evolved, they grew bigger brains with more neurones. To fit them all in, the grey matter had to fold itself into wrinkles. The neurones have connections with each other that run all over the brain. This makes up white matter. Inside the spinal cord is the central canal which contains the cerebrospinal fluid. This supplies the brain and spinal cord with nourishment.

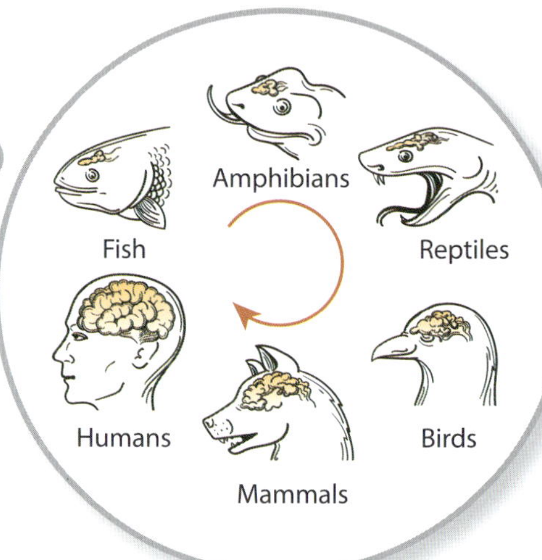

The grey matter does the processing, whereas the white matter provides communication

The Spinal Cord

The spinal cord starts at the back of the neck and extends to below the last rib. It takes information to the brain, and brings back orders from the brain to the body. It also takes care of the body's reflexes. The spinal cord is protected by the backbone which starts at the base of the skull and ends above the hip area.

The nerves work like telephone cables that carry messages from the organs to the brain and back. The organs in the head are directly connected to the brain, while those in the body connect to the spinal cord.

The brain and spinal cord govern the rest of the body. Nerves connect them all together

In Real Life

Look closely at the pictures of a walnut and brain. They look a little alike because like the brain, the walnut has folds and is divided into two halves. Additionally, eating walnuts is good for the brain. Walnuts are rich in minerals and lipids that help your nerves and brain function well.

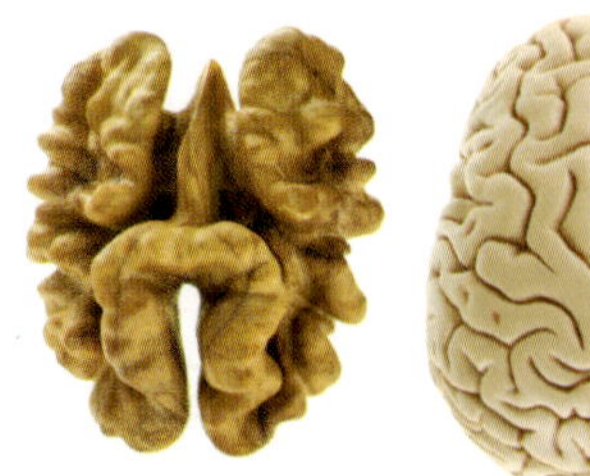

Some scientists claim that walnuts are the healthiest nuts of them all

Action and Reaction

Sense and Motor Organs

Imagine a pineapple cake in front of you. Your eyes give your brain a picture—it is big, round, and white. Your nose tells your brain that it smells of pineapples and cream. These are called stimuli. Your brain puts all the stimuli together and makes your hands reach out to grab a piece of the cake! As your fingers touch it, your brain knows that it is creamy on the outside and crumbly on the inside. You put it into your mouth, and your tongue tells you that it is sweet to taste. How did it all happen?

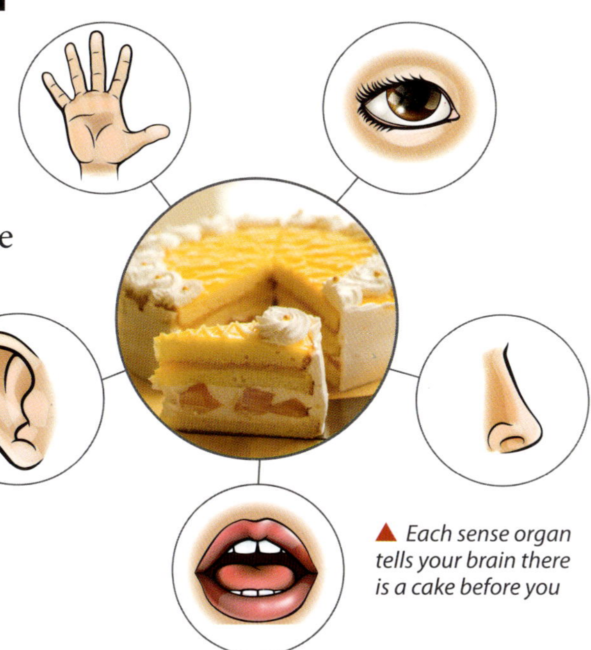

▲ *Each sense organ tells your brain there is a cake before you*

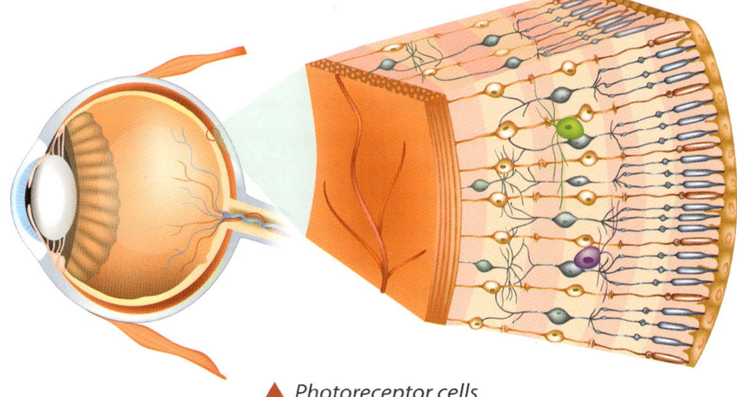

▲ *Photoreceptor cells*

Sense Organs

Each of our senses works in their own way. Chemoreceptors in the tongue and nose can sniff out the smallest amounts of chemicals in our food and air. Photoreceptors in the eyes pick up colours and light. Auditory receptors in the ears pick up the tiniest vibrations in the air. All of these turn into tiny electric signals that go through the nerves to the brain, which turns them into tastes and smells, sights and sounds.

Isn't It Amazing!

We cannot sense electricity by looking at a wire. But bees can detect the tiniest of electric currents from the flowers they visit. A bee can even sense whether another bee has already been to a flower that it is visiting. What's more, it comes to know whether the first bee took nectar from the flower. This saves the worker bee's time!

▲ *Bumblebees can sense electricity*

Motor Organs

When your sense organs tell your brain that there is a cake in front of you, your motor organs take action to grab a slice of the cake. Your brain sends signals through your nerves to your hand muscles, so they reach out for the cake. It also sends signals to your mouth and your voice box, so you can ask the grown-up in the room for permission to grab a slice!

▲ *Your motor organs act on the information they receive from the brain and the sense organs*

Fitting Together
Moving Pictures Sent on Radio Waves

The brain is divided into many parts. These include brain stem, little brain, midbrain, and higher brain. They work like a jigsaw puzzle where all the pieces make sense only if they fit together.

The diagram shows the parts of the inner brain

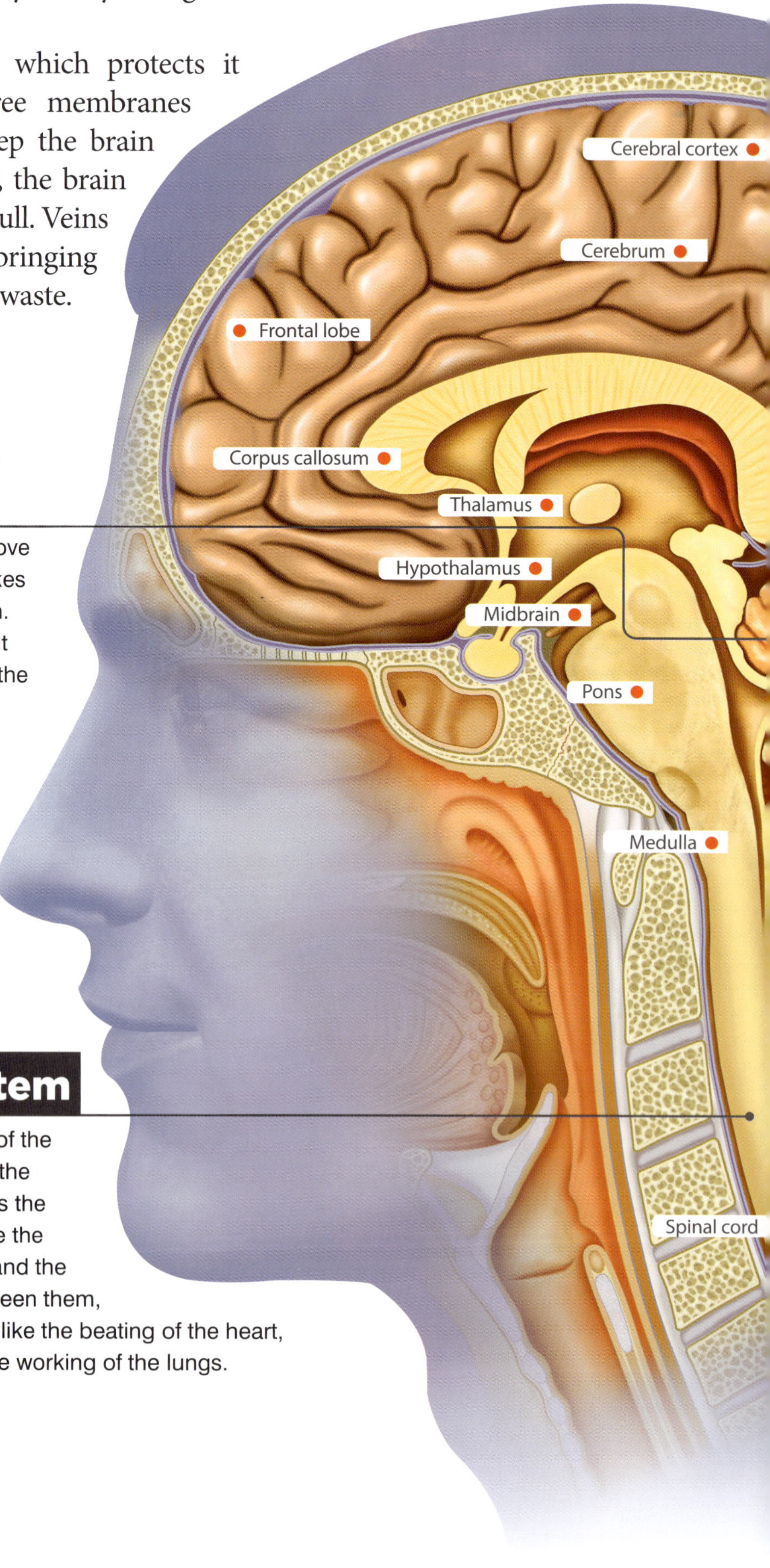

Cerebral cortex
Cerebrum
Frontal lobe
Corpus callosum
Thalamus
Hypothalamus
Midbrain
Pons
Medulla
Spinal cord

The brain is enclosed within the skull, which protects it from injury. It is also covered by three membranes called **meninges**. The meninges help keep the brain cushioned, so even if you hurt your head, the brain does not smash against the inside of your skull. Veins and arteries criss-cross the meninges, bringing nutrition to the brain, and taking away any waste.

The Little Brain

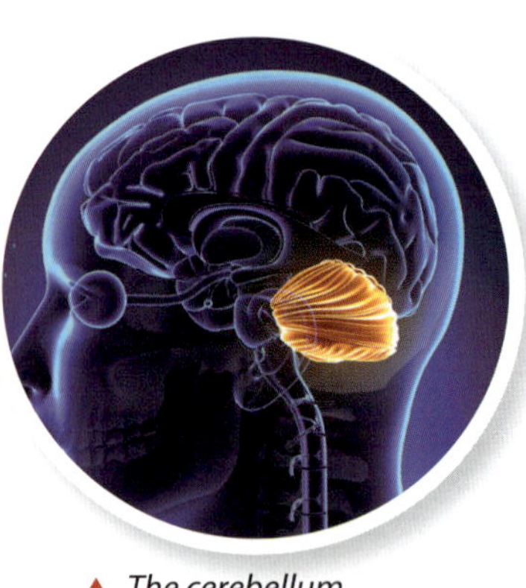

The cerebellum contains significantly more nerve cells than the cerebrum

This part of the brain is just above the back of your neck and makes up about one-tenth of the brain. Doctors call it the cerebellum. It acts like a checking centre for the brain, helping muscles correct themselves. For example, if something seems heavier than it looks, your cerebellum will get your muscles to put in more force. The little brain also helps you learn to avoid things that cause pain.

The Brain Stem

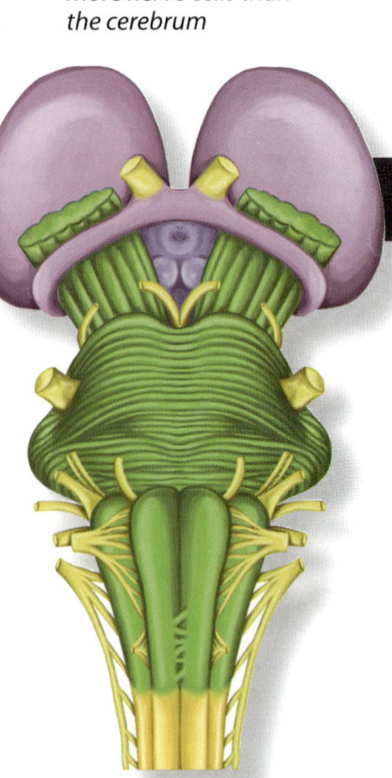

The brain stem is the first part of the brain. It is made of two parts—the **medulla oblongata** connects the spinal cord and the brain, while the pons connect the cerebellum and the cerebrum to the medulla. Between them, they control our daily activities like the beating of the heart, the circulation of blood, and the working of the lungs.

Structure of the human brain stem

The Higher Brain

The rest of the brain makes up the cerebrum. Most of what you see is the outer **cortex**, divided into two brain hemispheres (right and left) by the longitudinal fissure. The cortex is made up of four lobes, divided by inward folds called **sulci**. The outward folds of tissue that make up each lobe are called **gyri**. Each lobe has a function and also helps the other lobes.

The temporal lobe is responsible for hearing, memory, and learning. The occipital lobe is responsible for the function of sight. The parietal lobe is responsible for touch and movement. The frontal lobe is for thinking, acting, language, and personality.

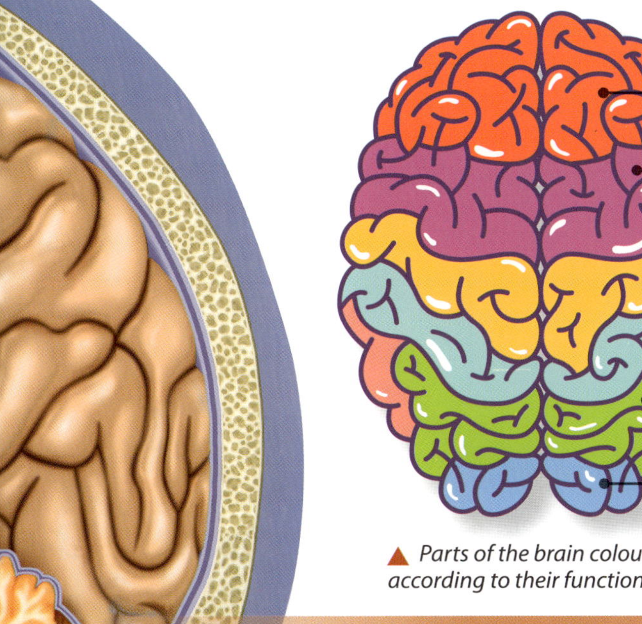

▲ *Parts of the brain coloured according to their functions*

In Real Life

A few decades ago, doctors would prescribe cutting off bits of the brain to cure moral deviancy and brain diseases like epilepsy. They called this lobotomy. More often than not, it left the patients worse than before. Often, after a lobotomy, a patient was unable to show emotion. Modern doctors think it was a cruel and useless practice.

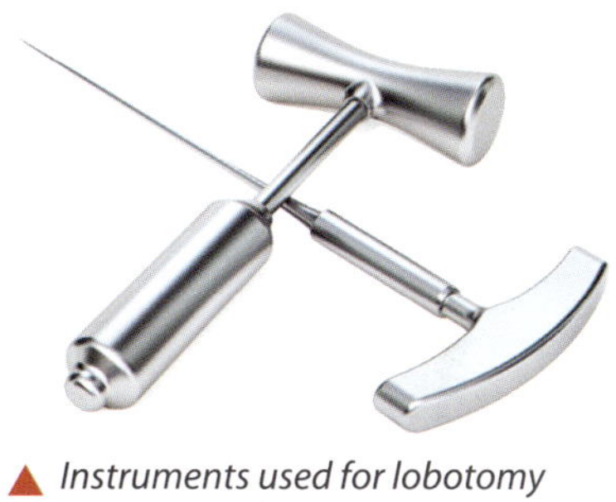

▲ *Instruments used for lobotomy*

The Inner Brain

The inner brain is made up of the **thalamus**, the **hypothalamus**, and the midbrain. The thalamus is the part that decides which stimuli to give attention to. This is the one that says, 'If you see a tree, do nothing. But if you see a lion, you better run!' The hypothalamus takes care of things like feeling hungry or thirsty, tired or fresh, sleepy or awake. The midbrain connects the senses, if you hear mum's voice, you immediately turn to see where she is.

Incredible Individuals

The procedure of lobotomy was developed by Antonio Egas Moniz but it was American neurologist Walter J. Freeman II, who popularised its use despite widespread criticism. He approached the media to promote the procedure, even giving it the name 'prefrontal lobotomy'. During his time, he performed around 3,500 lobotomies. Thankfully, he was banned from doing it after the last lobotomy, which happened in 1967

▲ *Antonio Egas Moniz*

▲ *Walter J. Freeman II*

Breaking News to the Brain

All our senses can be divided into two. There are senses that need to be felt, such as touch, taste, and smell. Then, there are senses like sight and sound that are not felt. You can also divide them into those that help you ascertain the direction of the sensory stimulus. The sense organs in the head communicate directly to the brain, while those in the rest of the body (mostly in your skin) communicate with the spinal cord. The sense of balance is unique as it does not depend on external stimuli.

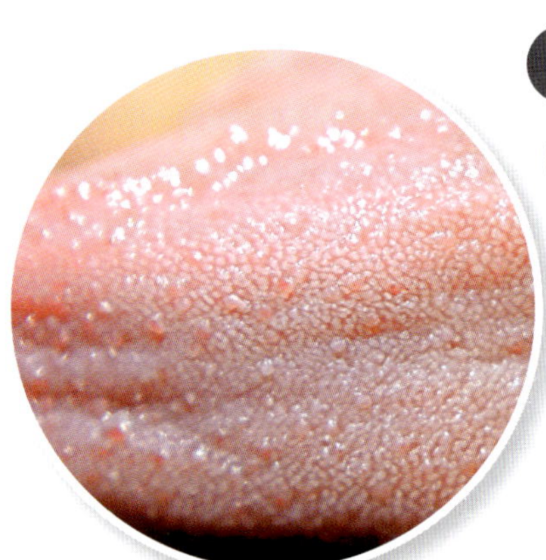

The tongue has around 3,000 to 10,000 taste buds

Taste

If you look at your tongue in the mirror, you will see that it is made up of hundreds of grainy bumps that doctors call **papillae**. Each papilla has a number of special nerves called **gustatory receptor cells** (GRCs). Different papillae have different kinds of GRCs for things that taste sweet, sour, bitter, or salty.

Smell

Deep inside the nose is an organ called the **olfactory bulb**, which gives us the sense of smell. It is made of thousands of **olfactory receptor cells** (ORCs), which catch chemicals in the air we breathe and tell us whether they are nice, like oranges, or pungent, like garlic. The nose also has pain receptors, which help pick up very strong smells that suggest that you may be exposed to a dangerous chemical like ammonia.

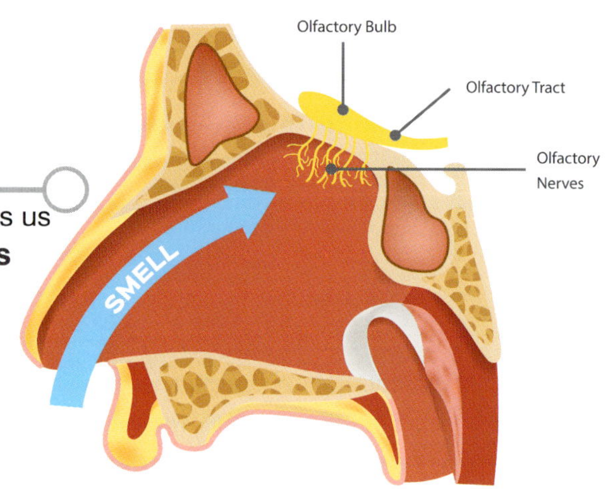

Cutaneous receptors

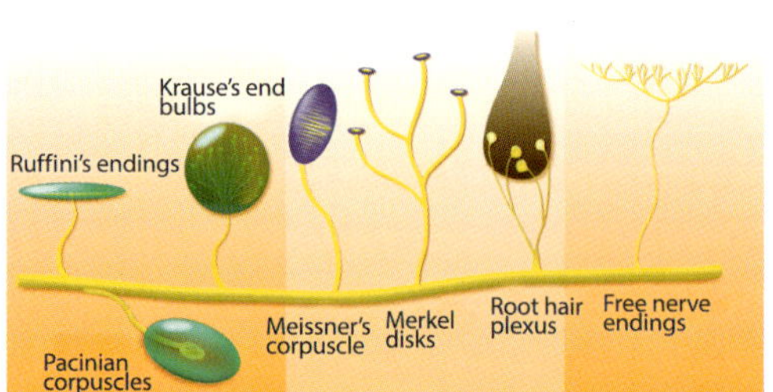

Touch, Heat, and Pain

Our skin is a giant sensory organ that can feel many things. Different kinds of nerve endings in the skin tell the brain and spinal cord different things. Each part of the skin is represented in the brain in a map called the homunculus.

Isn't It Amazing!

Butterflies have their taste buds on their feet!

Butterflies put their feet on the flower to quickly decide if it has nectar

In Real Life

A lot of what we think of as 'flavour', is actually both smell and taste together. The brain treats them together. That is why, when you have a cold and your ORCs are blocked due to a runny nose, food tastes odd even though your GRCs are not blocked.

Having a cold prevents you from tasting or smelling things properly

Sight and Sound

Our eyes and ears do not just see and hear. Since we have two of each, the brain can also ascertain where the stimuli are coming from, using the eyes and ears. So, if you hear something falling nearby, you know that you have to run in the opposite direction to save yourself. When you are playing cricket, your eyes tell you how far away the ball is from where you are standing.

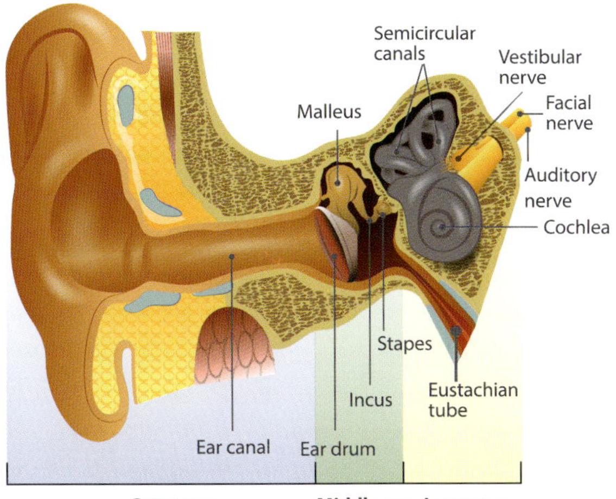

▲ *Your outer ear never stops growing throughout your lifetime*

Hearing

The ear 'hears' just like a telephone does. The outer ear acts like a microphone, collecting sound from around you. These sounds go through your ear canal to the ear drum. This is a tiny stretch of skin that vibrates, just like a telephone diaphragm. The tiny ear bones pick these vibrations and take them to the cochlea. The cochlea is coiled like a snail's shell and breaks what you are hearing into separate sounds, so you can hear even if separate people are calling for you at the same time.

The brain can hear the same sound through both ears, so it can figure out whether the sound is coming from the left, the right, the front, or behind you.

Balance

Do you see the semi-circular canals in the figure above? These have nothing to do with hearing, but tell your brain about your body's balance. They are filled with a jelly-like substance and have tiny hairs inside them. If you are wobbling without balance, the jelly moves and the hairs pick up on this movement and communicate to the brain. The brain then sends a message to the relevant muscles to steady yourself, even without you realising it.

Sight

The human eye works just like a camera. The pupil and iris in the front of your eye act like shutters that close and open to let in light. The lens concentrates light just like the lens of a camera. Light finally lands on the retina, which has special **photoreceptor cells** (PRCs). They are of two types—**rod cells**, used for night vision; and **cone cells** that work better in daylight. The cone cells sense only three colours of light: red, green, and blue.

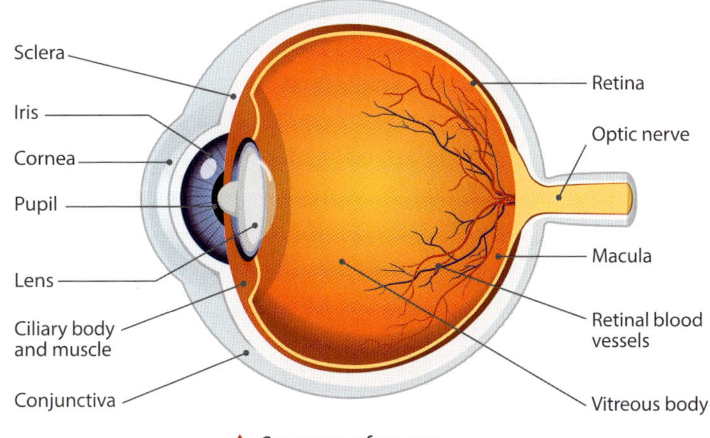

▲ *Structure of an eye*

▲ *Humans have a maximum horizontal field of 200° with both eyes*

Binocular Vision

Both your eyes tell the brain slightly different things. What is right in front of you is the same, but the left eye sees things to your left, and the right eye sees things to your right. Using these differences, the brain can make a 3D map of the world in front of you, so not only do you see things, but you also know how far they are.

Nerves and Neurones
How They Work

The neurone is a nerve cell that forms the basic unit of the nervous system. Unlike other cells, neurones can stretch for several inches along the body. They clump together in the brain to make nuclei. In the rest of the body, neurones clump together to make **ganglia**. Bundles of neurones that interact with the same organ are called nerves.

▼ *The diagram provides a closer look at the nerves, neurones and axons of the body*

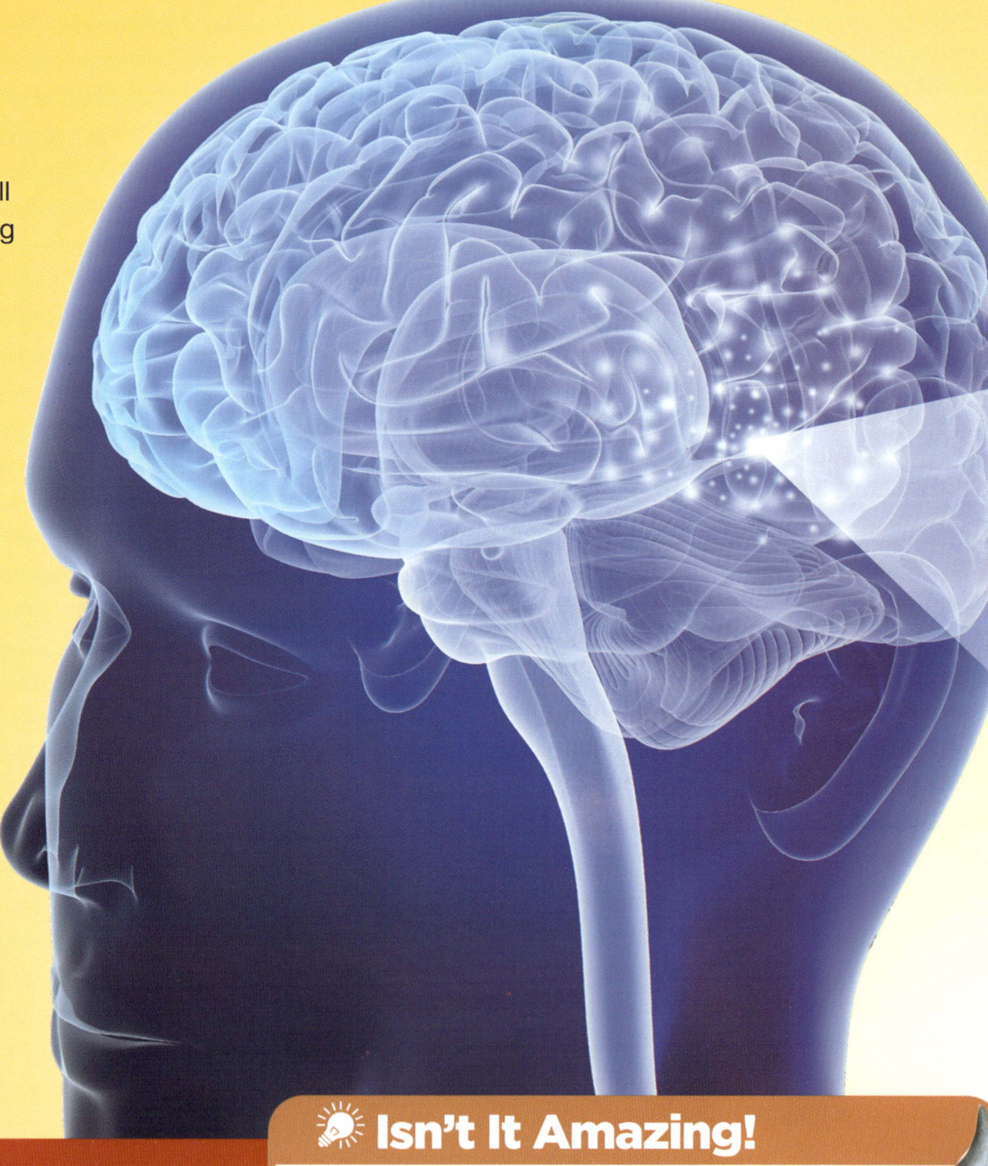

Cell Body

Most of the neurone's mass sits in the cell body. This has lots of mitochondria, giving the neurone all the energy it requires.

Dendrites

Dendrites look like plant roots and are parts of the neurone that touch other body cells or neurones. They gather information from the sensory cells and pass them onto the axon at the other end.

Axon

This is the longest and most important part of a neurone. It carries news from the dendrites to the next neurone over long distances. This information is carried in the form of tiny amounts of electricity called an action potential. **Axons** in the brain make up its white matter and are not covered by myelin sheaths.

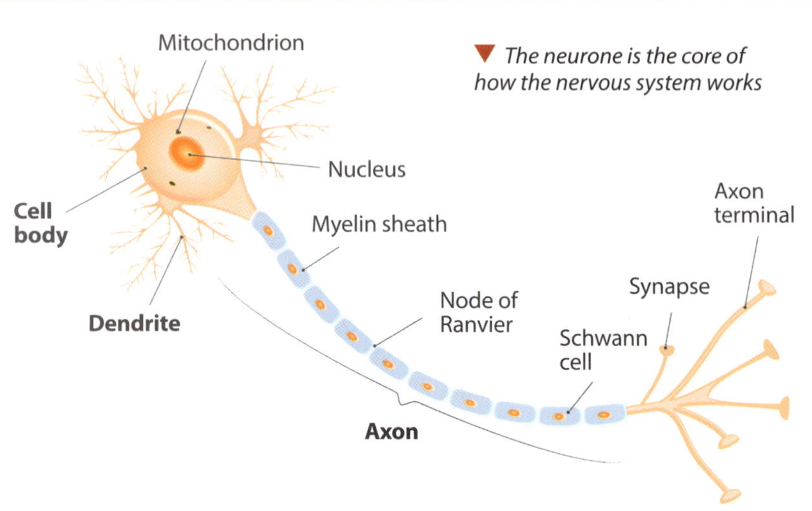

▼ *The neurone is the core of how the nervous system works*

Isn't It Amazing!

A blue whale can grow up to 34 metres long. Some of the neurones in its spinal cord run all the way to the tip of its tail. At nearly 30 metres, this makes the blue whale's spinal neurones the longest single cells in the world!

▲ *Blue whales have the longest neurones in the world*

Myelin Sheath

Like the plastic insulation that protects electric cables, the **myelin sheath** protects the axon, so that the information does not leak. It is made of a material called myelin, which is made by the **Schwann cells**. The glial cells wrap themselves around the axon, making an extra cover. We call the gaps between Schwann cells the **Nodes of Ranvier**. These allow the neurone to recharge itself.

Synapses

A synapse is where one neurone meets another to pass on the information. A neurone will have synapses with dozens of others. They also act like switches, slowing down or speeding up news travelling through the nerves. The brain makes new synapses between different neurones all the time, and that is how we make new memories and learn new things.

Action Potential

This is a tiny electric current that passes along the wall of the axon. When there is no message to be passed along the nerve, there is sodium (Na^+) outside the neurone and potassium (K^+) and chloride (Cl^-) inside it. This is the resting potential. When there is a message to be passed, Na^+ rushes in and K^+ rushes out, creating the action potential. After the message has passed, the two slowly switch places again, bringing the neurone back to normal.

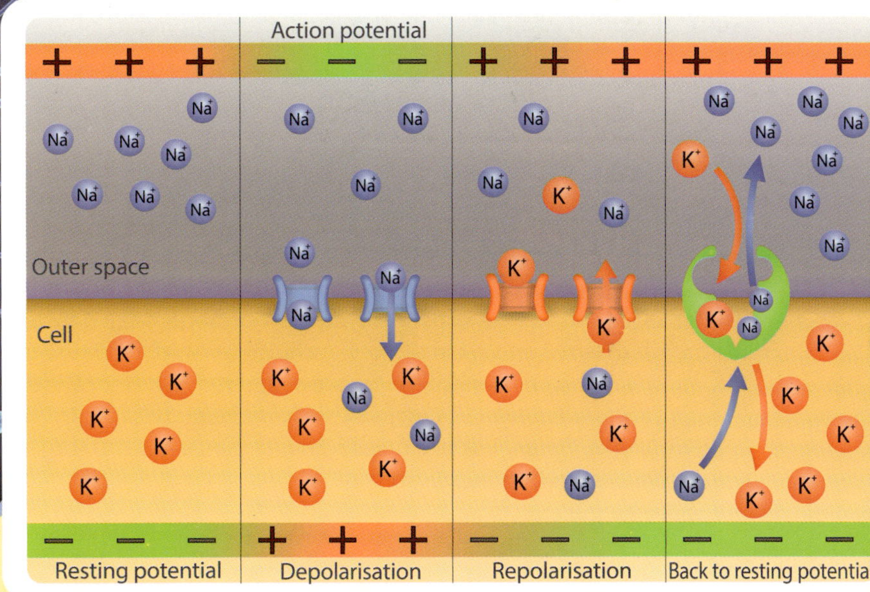

▲ *Neurones take messages to and from the brain through action potentials*

Mirror Neurones

These are neurones found in many parts of the brain. They help to bridge some of the senses with action. Scientists conducted experiments in which they found that these neurones become active when we see or hear something that is familiar to us. For example, these become active in ballet dancers' brains when they see other ballet dancers. These neurones help us learn things by seeing or hearing them, and also in expressing feelings like sympathy.

Incredible Individuals

Santiago Ramón y Cajal (1852–1934) was a Spanish neuroscientist who discovered that the neurone was the basis of the nerve cell. He developed techniques to 'stain' individual neurones and follow their path across the nerves. Many of the drawings he made are still used by medical teachers.

▲ *Santiago Ramón y Cajal was the first Spaniard to win a scientific Nobel Prize*

Your Body's Other Brain

The spinal cord takes messages from the organs of the body to the brain and back from it to the organs. It also takes messages to the muscles from the higher brain, but it does a lot more than acting like a courier. It also governs the sympathetic and parasympathetic nervous systems, which do not depend on the brain. It further handles automatic reactions of your body that the brain does not get told about, like your reflexes. That is why it acts as your body's second brain. Any damage to the spinal cord makes you unable to move some parts of the body, which doctors call paralysis.

Spinal Cord

The spinal cord runs entirely inside your backbone. Unlike the brain, the white matter is outside, and the grey matter is inside. The white matter is made of the axons and dendrites of sensory neurones coming in from the various organs and the skin, motor neurones going out to the organs and the skin, and neurones travelling through the length of the spinal cord, connecting to the brain stem. The grey matter is made of the cell bodies of these neurones, and small connecting neurones called **interneurones**.

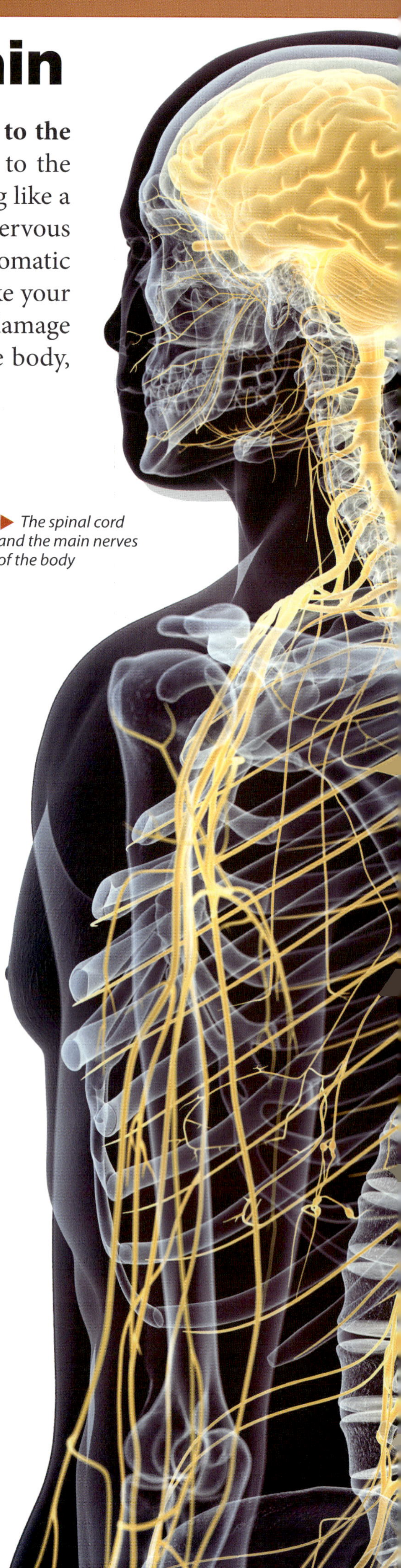

▶ *The spinal cord and the main nerves of the body*

In Real Life

Does the octopus have a backbone? No! That is precisely why it is called an invertebrate marine animal. This feature, along with them being extremely intelligent and curious creatures, makes octopi incredible escape artists. For instance, Sid, an octopus who was in captivity in New Zealand, escaped his tank multiple times before finally being released by the aquarium officials as they were fed up.

▲ *When bored, octopi are known to amuse themselves—from juggling hermit crabs to short circuiting aquarium lights*

Reflexes

Reflexes are reactions of the body that do not require you to think. These are actions you take out of fear, hunger, pain, or anything else that requires you to act quickly. During a reflex, the sensory neurone connects with an interneurone through a synapse, which connects to a matching motor neurone. For example, some neurones from your fingers that can sense very hot things are connected to neurones in your forearm that pull the hand away. So, when you touch a pie right out of the oven, you feel the heat and pull your hand away, dropping the pie. Your eye sees the mess and your hands get to work cleaning it up. That is a second reflex! But it is the one that goes through the brain.

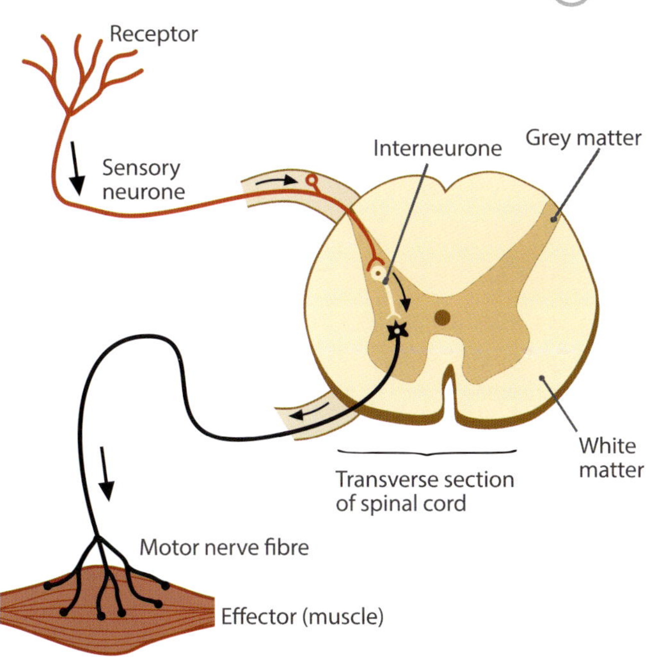

▲ *Spinal reflexes allow you to act without waiting for the brain to decide*

There are two kinds of reflexes. The somatic reflexes connect the sensory organs with skeletal (voluntary) muscles. These often act as part of the sympathetic nervous system, which helps you to deal with sudden situations. You can train yourself to stop these reflexes.

Visceral reflexes connect the sensory organs to internal (involuntary) muscles and you cannot control them. These are often part of the parasympathetic nervous system, such as feeling hungry after seeing a cake.

Keeping Your Two Brains Safe

The brain and spinal cord are not supplied with blood directly. Instead, the cerebrospinal fluid keeps them nourished. Blood capillaries in the brain are surrounded by tiny cells called glia. The glia makes up the blood-brain barrier and makes sure that nothing other than water and minerals can leave the capillaries and enter the cerebrospinal fluid. Proteins, large fats, and germs cannot cross this barrier. Glucose is moved into the brain by active transport.

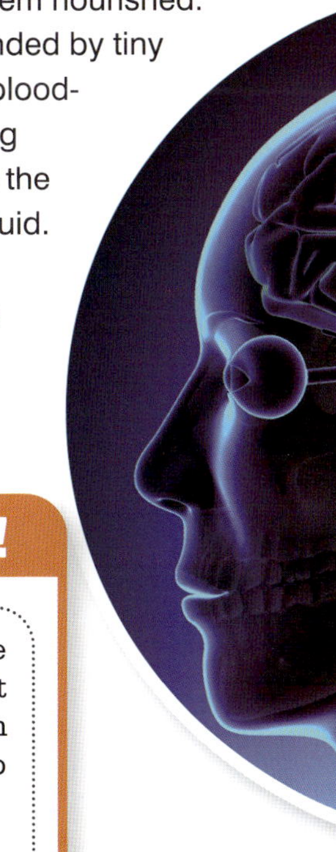

Isn't It Amazing!

Chameleon tongues have some of the fastest reflexes in the world. Once it has seen its prey, a chameleon can shoot its tongue out at 2.6 kmps to catch it.

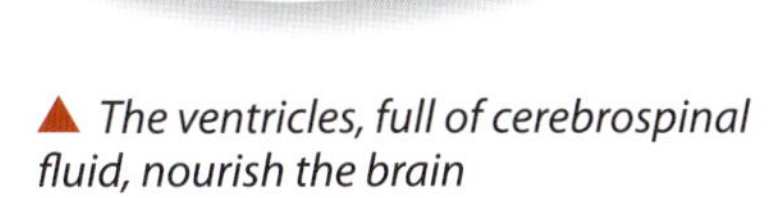

▲ *The ventricles, full of cerebrospinal fluid, nourish the brain*

◀ *Chameleons have some of the fastest reflexes in the world*

The Pituitary and Pineal Glands

When we are tired, we go to sleep. When we go out in the Sun, our body makes more **melanin**, giving us a tan. When we have been sufficiently rested, we wake up. How does our body know all this? This is where the endocrine system comes in. The organs of this system are called glands and make different **hormones**, which are chemicals that travel through blood and tell the other organs of your body what to do. The glands of the body, such as the liver, pancreas, thyroid, and adrenal gland are, in turn, governed by the pituitary gland. You can call it our body's chemical brain.

The Pituitary Gland

The **pituitary gland** makes hormones that control how other glands work. In turn, the pituitary is controlled by the hypothalamus, which contains centres for controlling hunger, thirst, body temperature, tiredness, etc. It makes the:

Growth hormone, which makes your body grow.

Prolactin, which helps a woman's breasts to make milk for her baby.

Thyroid-stimulating hormones, which make the thyroid gland release the thyroid hormone.

Adrenocorticotropic hormones, which make the adrenal gland release adrenaline.

Antidiuretic hormones, which make the kidneys take up water filtered from the blood.

Melanocyte-stimulating hormones, which make the melanocytes in the skin make melanin, which protects you from the UV rays of the sun.

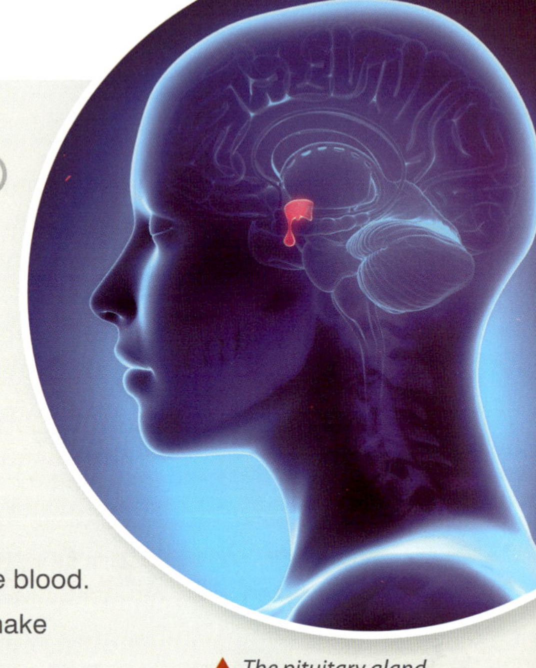

▲ *The pituitary gland is called the master gland of the endocrine system*

Pineal Gland

This is the brain's other gland, behind the hypothalamus. It makes a hormone called melatonin, which acts on the hypothalamus and tells it to go to sleep. You can also get melatonin from fish, eggs, nuts, and mushrooms. The pineal gland makes melatonin all through the day and releases it into the blood as daylight fades. When light falls on you, the brain and intestines make serotonin, which tell the brain to wake up. Over time, the cycle of sleep and waking becomes regular, and less dependent on light. This is called the circadian rhythm.

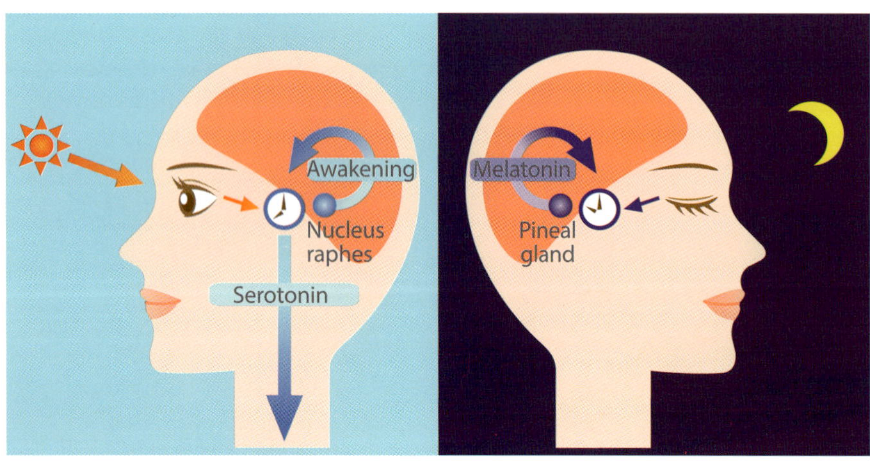

▲ *Melatonin makes you sleep; serotonin wakes you up*

Isn't It Amazing!

In many birds, reptiles, and mammals, the pineal gland is very important. The pineal gland helps these animals to know the right season to mate and reproduce. Removing this gland makes them unable to breed in the right season. In some species, the pineal gland can also detect light and make melatonin accordingly. So it is sometimes called the 'third eye'.

Improve Your Mental Health

Though we still do not have answers to all the illnesses of the nervous system, indulging in activities that help you unwind can be very fruitful for maintaining a good mental health. This could include anything: from the use of colours and paints to bring your ideas to life, to playing with your pup. Science also backs the use of varied forms of therapy to diagnose, as well as treat mental health conditions in children.

▲ *Psychologists watch children play and figure out signs related to their mental health*

Play Therapy

Play forms a crucial part of how you make sense of your world, as a child. It is also a means for children to express their feelings. Psychologists often use play therapy to aid children in processing complicated emotions such as grief resulting from loss of a parent or caregiver. Play therapy is also great to help you learn desirable behaviours and unlearn the faulty ones.

Art Therapy

It is not always easy to express your complicated emotions. In order to do so, you can draw art. Psychologists sometimes suggest drawing, painting, sculpting, clay modelling, etc. for children with depression, anxiety disorders, and so on. It is also used with children diagnosed with ADHD as well as low IQ.

▲ *Art therapy helps you explore your inner 'experience'*

Animal-assisted Therapy

Animal-assisted therapists have specially trained animals, who are friendly and warm, so you enjoy playing with them, and feel less worried or frightened about things. It facilitates a child in communicating their complicated emotions better with the therapist. So, when you are feeling bad, give your dog or cat a big fluffy hug!

Isn't It Amazing!

Catching a flight can be stressful. Before the flight, you are probably hoping you make it in time past security check or long queues. Airports around the world now have new kind of employees—dogs and cats who help passengers calm down and enjoy their wait, known as Emotional Support Animals (ESAs).

◀ *Cats, dogs, horses, bunnies, and even goats help with animal-assisted therapy*

Inside Your Heart

Let us try and understand why the heart is so important. When a heart surgeon opens the heart, they see four chambers, separated by the inner heart wall called septum and valves. There are two on the upper, broad part called the base. These chambers are the **atria** (singular: atrium) or auricles, one to the left and one to the right. The other two chambers are the left and right **ventricles**, which are in the apex, the narrower, lower part of the heart.

There are valves between the chambers, just like there are doors between the rooms of your house. They work like gates, stopping blood from leaking out of a heart chamber once it has been filled. This makes the heart's work smooth, and also maintains your blood pressure.

Superior vena cava: It brings deoxygenated blood from the head and upper organs

Right atrium: This is the first chamber of the heart to receive blood from all organs, except the lungs. It has thinner walls than the rest and mixes the blood from the two vena cavas

Tricuspid valve: This valve lets in blood from the right atrium to the right ventricle

Right ventricle: Here blood collects till the chamber is full, and is then pumped into the pulmonary artery, through the pulmonary valve

Inferior vena cava: It brings deoxygenated blood from the lower organs of the body. Both vena cava join up and pour blood into the heart

Isn't It Amazing!

The left ventricle is the strongest and largest chamber of the heart, sending blood to all parts of the body. It pumps five times as much blood as the right ventricle.

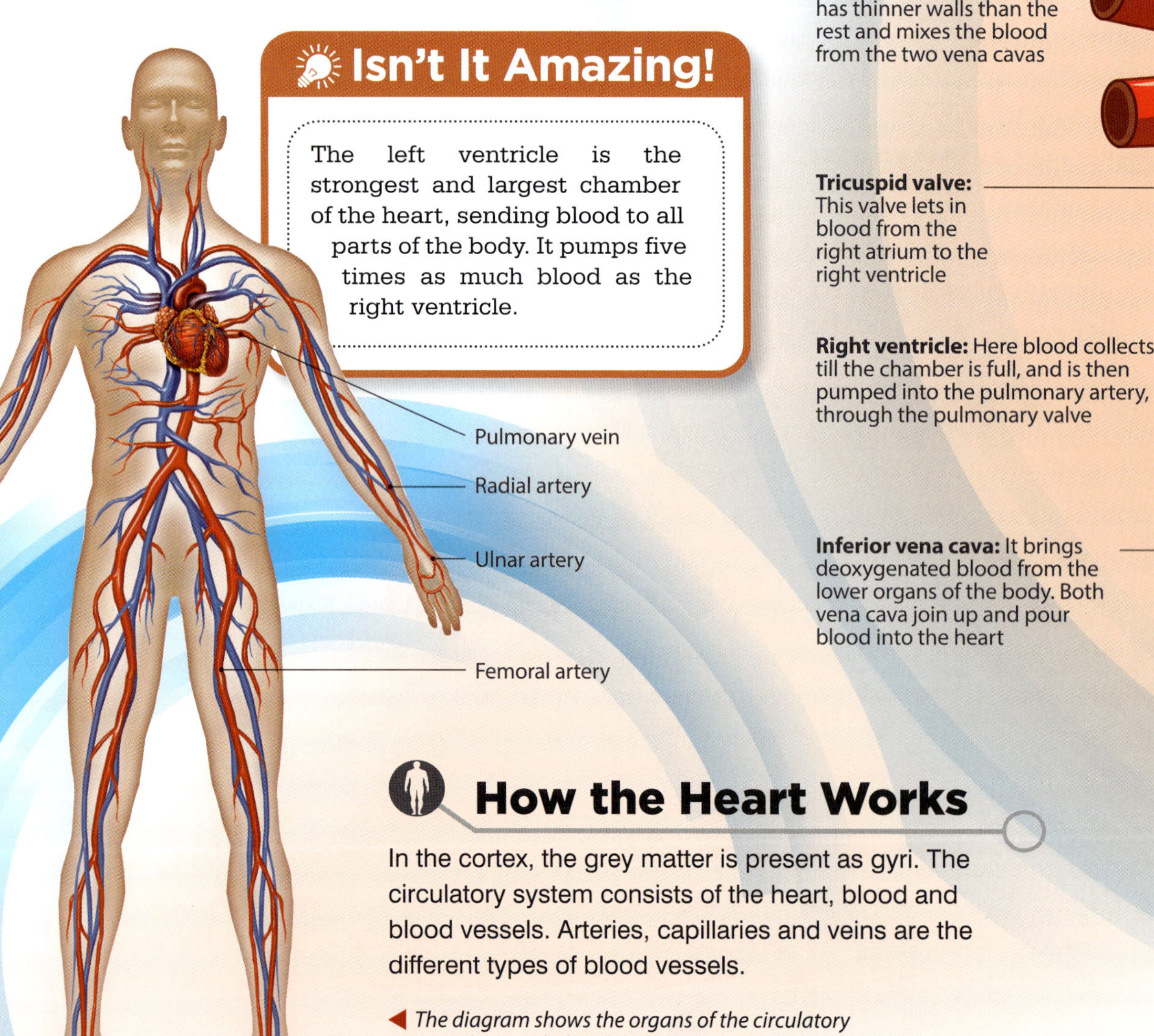

How the Heart Works

In the cortex, the grey matter is present as gyri. The circulatory system consists of the heart, blood and blood vessels. Arteries, capillaries and veins are the different types of blood vessels.

The diagram shows the organs of the circulatory system along with their labels

Circulatory Cycle

Every drop of blood finishes one cycle when it goes out of the heart carrying oxygen, through the organs, back to the heart, into the lungs and back from the lungs to the heart, where it receives fresh oxygen. This cycle is kept going by the cardiac cycle.

Aorta: This is the body's biggest artery and takes blood to all the organs. It loops around the heart and breaks into two—one going to the head and one to the lower body

Pulmonary artery: This comes out of the heart and branches into two. It carries deoxygenated blood to the left and right lungs, where the carbon dioxide is released into air and blood becomes oxygenated

Pulmonary veins: These bring back blood, freshly oxygenated from the lungs to the left atrium

Left atrium: This has a thicker wall than the right atrium. It collects blood coming from the lungs

Mitral valve: This valve opens when the left atrium is full and lets blood into the left ventricle

Left ventricle: This pumps out the oxygenated blood to the rest of the body through the aorta. It is plugged by the aortic valve

▲ *The chambers of the heart are joined by valves*

Pumping of the Heart (Cardiac Cycle)

The heart never rests. But it has a quiet diastolic phase when the ventricles are being filled with blood. Once they are full, the heart enters the active systolic phase. The heart muscles contract together, and blood is pumped out with force into the aorta and pulmonary artery. Together these make up one heartbeat. The force with which your heart pumps decides the pressure with which blood flows.

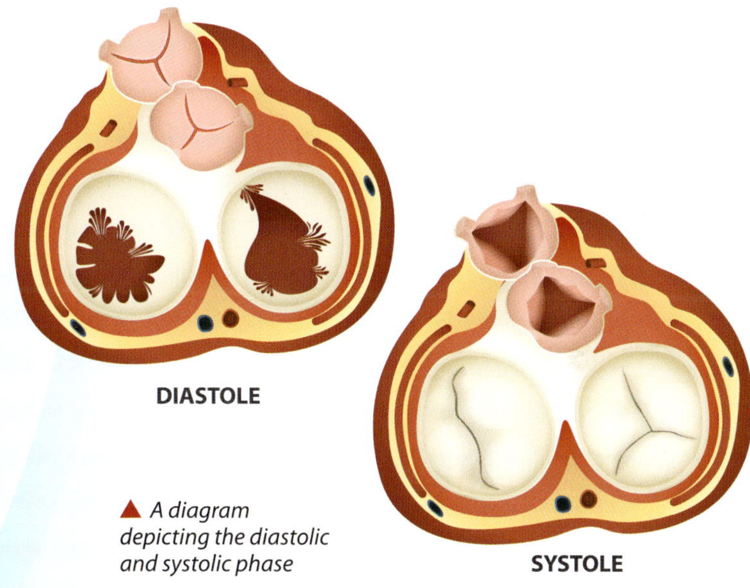

▲ *A diagram depicting the diastolic and systolic phase*

Evolution of the Heart

Unlike human beings (who are **vertebrates**), invertebrates have no blood to transport oxygen or nutrients. Air reaches tissues directly through spiracles, while nutrients are absorbed through a fluid called haemolymph. The heart, as an organ, originated in fish—with two chambers: an atrium to receive blood from the gills, and a ventricle to distribute the blood to tissues through arteries. After lungs evolved in amphibians and reptiles, the atrium split into two—one for deoxygenated blood from the tissues, and one for oxygenated blood from the lungs. But the two get mixed in a single ventricle. In crocodiles, birds, and mammals the heart finally becomes four-chambered, and deoxygenated blood is completely separated from oxygenated blood.

Arteries & Veins

Did you know that within a minute, blood from your heart will have reached every cell of your body? That is because our body has an efficient transport system made of arteries and veins, together called blood vessels. Blood flows in the arteries away from the heart towards the organs. The oxygenated blood in them makes them look red. In the veins, blood flows towards the heart from the organs. The deoxygenated blood in them makes them look blue.

What Do Arteries Do?

Oxygen is very important for your body's cells as they need it for making energy from the food you eat. As they keep using it, they need more. It is the arteries that bring them the oxygen from the heart, which it gets from the lungs. Other than oxygen, your arteries bring hormones from various glands in the body. Hormones help communicate to the cells what to do and when. Arteries also bring nutrients like glucose and amino acids from the intestines, which digest the food you eat. The pulmonary artery is the only one that's different, as it carries deoxygenated blood from the heart to the lungs for oxygenation.

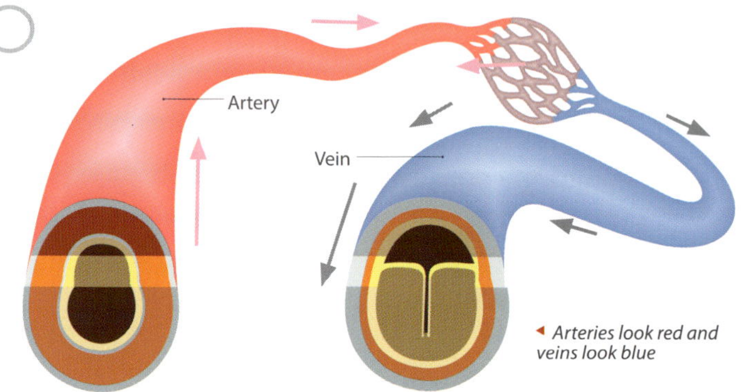

Arteries look red and veins look blue

What Do Veins Do?

Carbon dioxide is what is left once food is turned into energy (or used for making proteins needed by your tissues). It must be removed from the body so that fresh oxygen can be brought in. This job is done by the veins. They also carry away other wastes from the cells, like urea. Two veins are different from the others. The **pulmonary veins** bring blood rich in oxygen from the lungs to the heart. The hepatic portal vein takes blood with glucose from the intestine to the liver, where the extra glucose is stored.

How are Arteries and Veins Made?

Like your heart, your arteries and veins also have three layers wrapped around each other. The outermost layer is the 'tunica adventitia' which is made of loosely bound cells called 'connective tissue'. Next to it is the 'tunica media', which has tiny muscles that help push the blood along. It is thicker in arteries and thinner in the veins. Innermost is the 'tunica intima', made of a wall of cells that are tightly bound to ensure no leaks occur.

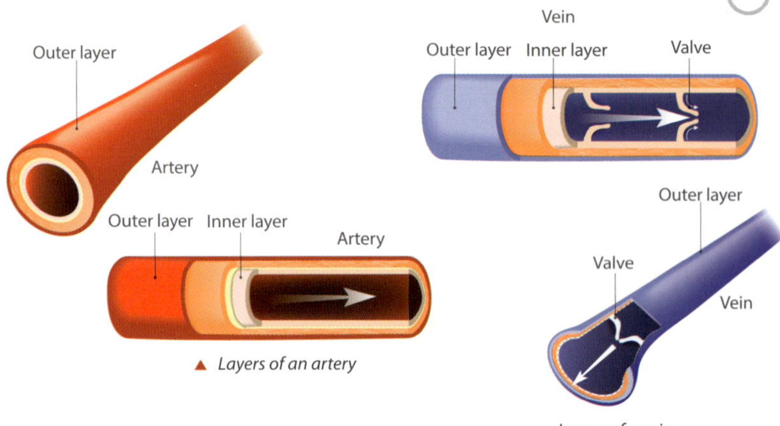

Layers of an artery

Layers of a vein

Deep & Superficial Veins

The inferior vena cava is so called that because it lies posterior to the heart. It is the largest vein in the body, collecting deoxygenated blood from all the organs. The veins that bring blood to it are called deep veins, because they are deep inside your body. The veins on your skin are called superficial veins. You can see one stick out of the insides of your elbows. These veins connect to the deep veins.

Feeding the Organs

Your aorta starts from the left ventricle of the heart, and gives out branches or arteries that go to each organ. Inside the organ, they branch into smaller arterioles. As they go deeper into the organ, they branch out into really thin blood vessels called capillaries. Did you know that the heart has its own artery that supplies blood to each of the heart's muscles?

Isn't It Amazing!

Look at one of your hair strands. Some blood vessels are extremely small; that the smallest blood vessels measure 5 mm, that is less than one-third of a hair strand!

Human hair under a microscope

Capillaries

Capillaries act like both arteries and veins. They do not just supply oxygen and nutrients, but also pick up waste and carbon dioxide from the organs including the heart. They join to form venules, which come out of the organ to form the main vein, which in turn joins the vena cava. You can see a network of capillaries if you look at your eye in a mirror by gently pulling down an eyelid.

Precapillaries are located between the capillaries and the smallest arterioles or arteries. These are considered to be intermediate vessels. Precapillaries control how the capillaries are emptied and filled. They have muscle fibres, unlike the capillaries.

Types of Capillary

Your body has capillaries of three types. They are the continuous capillary, the fenestrated capillary, and the sinusoid or discontinuous capillary.

Continuous Capillary

These capillaries have thin gaps between the cells that make up its walls. This lets the fluid part of blood go out into the tissue, and come back in. This kind is found in the lungs, muscles, and nervous system.

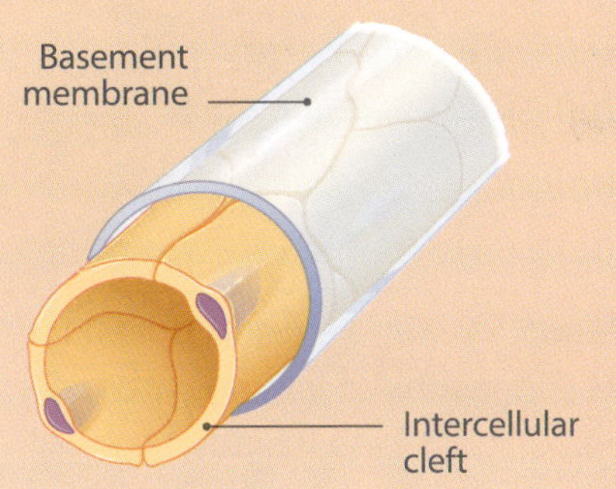

Fenestrated Capillary

The fenestrated capillaries have little holes (like a shower head) that let blood go out and come in. This kind runs in the kidneys, intestines, and glands.

Capillaries are usually 10 μm in size

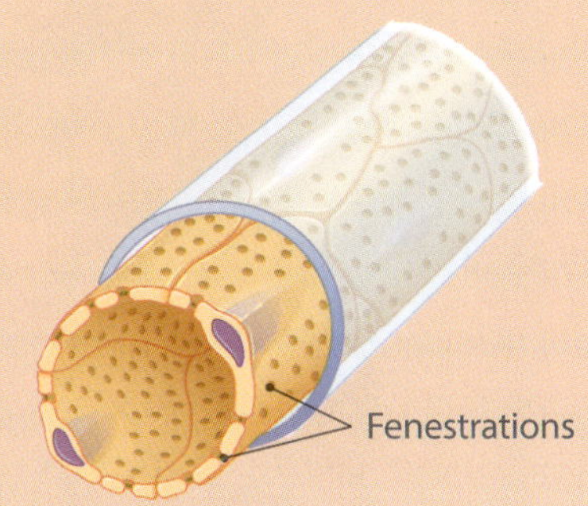

Sinusoid Capillary

Lastly, the sinusoid or discontinuous capillaries have really big gaps that allow blood cells to get into the tissue. You see them in the liver; spleen, where damaged blood cells are destroyed; and bone marrow, where new blood cells are born.

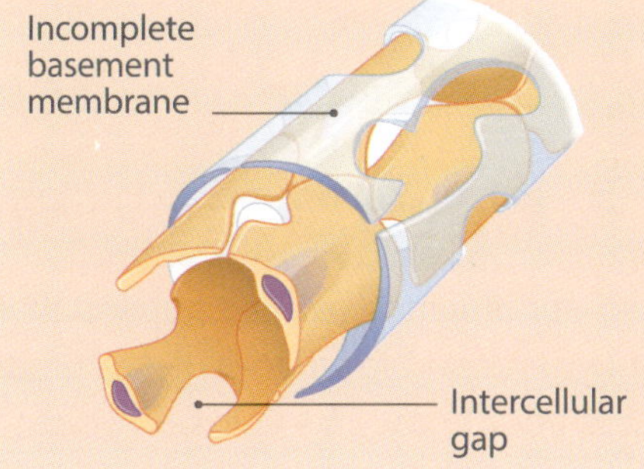

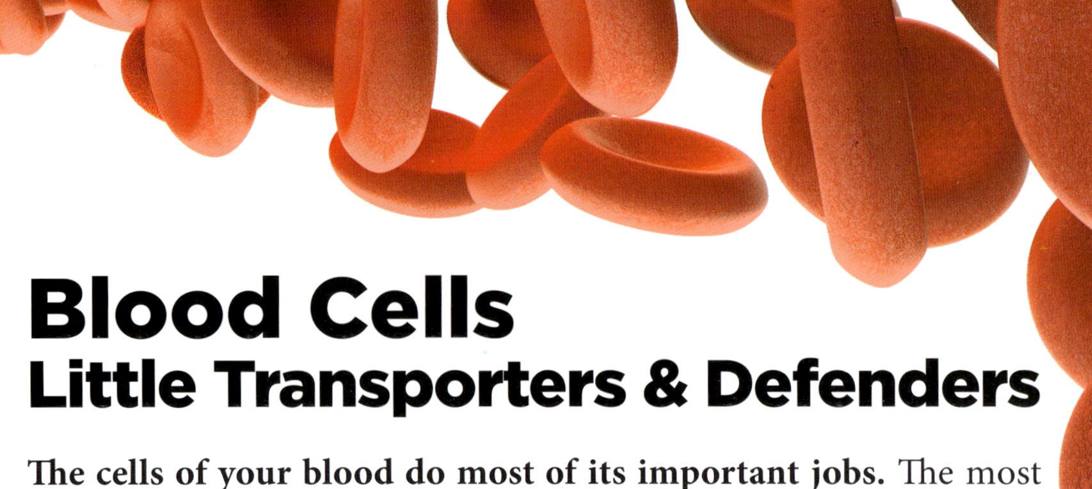

Blood Cells
Little Transporters & Defenders

The cells of your blood do most of its important jobs. The most in number are the red blood cells (RBCs) or erythrocytes. They give your blood its colour and carry oxygen from the lungs to the organs through the heart. Each drop of blood has about 5 million RBCs. On the other hand, white blood cells (WBCs) or leucocytes make up less than 1 per cent of blood. They help defend the body by swallowing any germs that may enter your body. Both RBCs and WBCs are made in the bone marrow, and at the end of their lives, are destroyed in the spleen and liver.

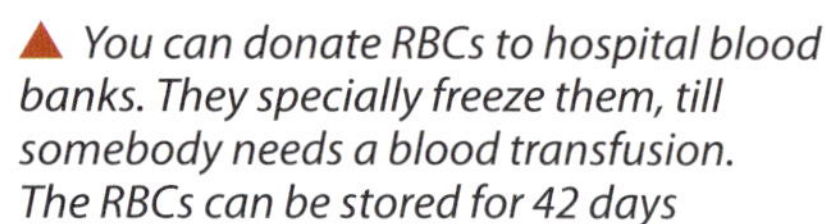

▲ *You can donate RBCs to hospital blood banks. They specially freeze them, till somebody needs a blood transfusion. The RBCs can be stored for 42 days*

Isn't It Amazing!

If there are 5 million RBCs in one drop of blood, how many are there in the whole body? 30 trillion! Every fourth cell in your body is an RBC.

Red Blood Cells

Under a microscope, RBCs look like disc-shaped doughnuts (without the hole), making up 40 per cent of your blood. They travel through your body and live for about four months. Each second, 2 million new RBCs are born, and 2 million die. They are made from 'haematopoietic stem cells' in the bone marrow; it takes about a week for them to turn into 'mature' RBCs. Then they enter the blood through the 'sinusoid capillaries' to get on with their jobs!

What Do Red Blood Cells Do?

RBCs have a special protein in them called **haemoglobin**, which contains iron. The iron makes haemoglobin, RBCs and therefore your blood red in colour. In the lungs, each unit of haemoglobin binds four molecules of oxygen and carries them to the rest of the body's tissues, which are hungrily waiting for them. They also take carbon dioxide from the tissues to the lungs, where it is released into the air.

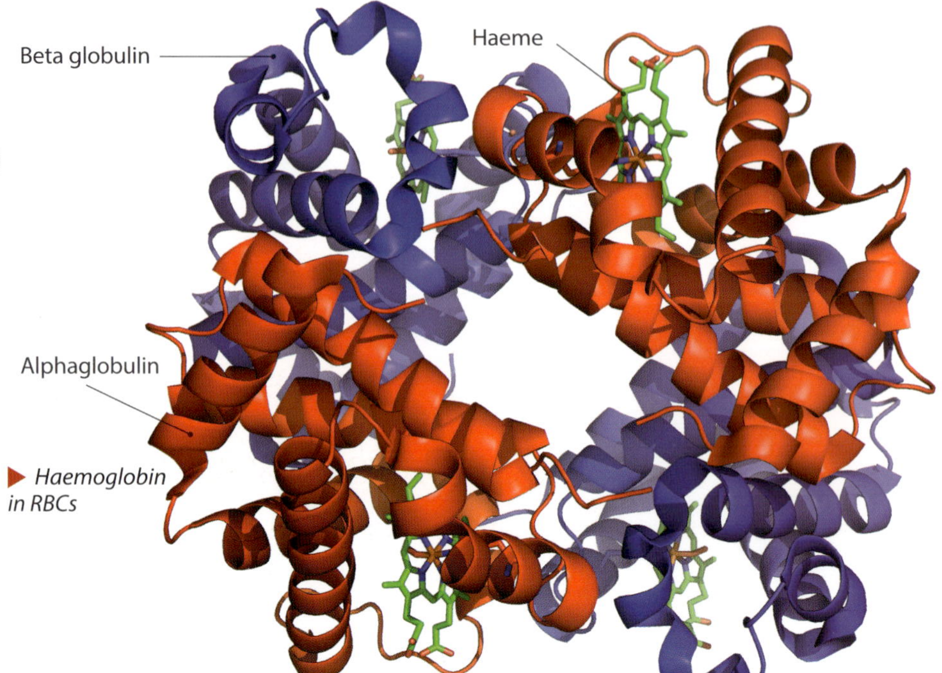

▶ *Haemoglobin in RBCs*

White Blood Cells

WBCs come in many types—**macrophages**, lymphocytes, neutrophils, basophils, and eosinophils. Each type of cell does something different in protecting the body from germs and allergens. They are actually a part of the immune system, where the body fights diseases and infections. Some WBCs live only for a day, while some may live for several years, remembering a previous infection just in case you are infected again. WBCs cannot be donated easily as it is hard to extract them from blood.

▲ *White Blood Cell (WBC)*

How Do White Blood Cells Work?

The different types of WBCs work together like a team to fight germs. Along with the lymphatic system, they make up the immune system. Here is what each WBC does.

Types of White Blood Cell

Neutrophil: Neutrophils go first. They attack most of the harmful germs and send signals to other WBCs to help.

Basophil: Basophils cause inflammation if an allergy-causing substance such as pollen or nuts enters the body.

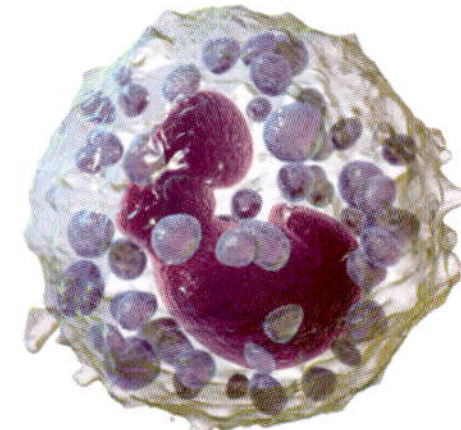

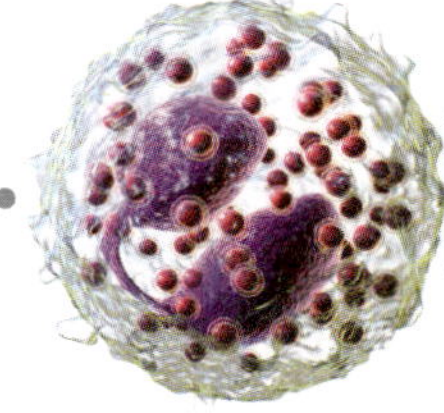

Eosinophil: Eosinophils kill bacteria and parasites in our body and clean up dead cells.

Lymphocyte: Lymphocytes come in many types themselves, making antibodies and other biochemical weapons.

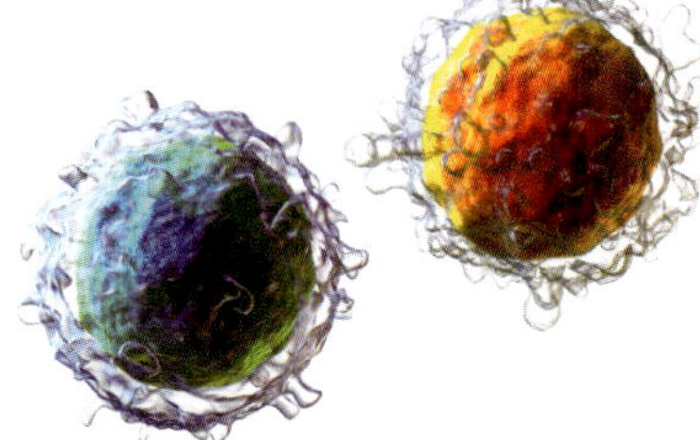

Monocyte: Monocytes transform into macrophages, which swallow infected cells, germs, and dead cells.

Isn't It Amazing!

The cells also interact with each other through special biochemicals called cytokines and interleukins.

Lymphatic System
Your Body's Janitor

The lymphatic system is similar to the circulatory system, but lymph vessels are thinner than blood vessels. They end in two major lymph ducts, the right lymphatic duct and the much larger thoracic duct. Lymph only flows in one direction—back to the heart. For this, it has to travel upwards, as the two lymph ducts meet the circulatory system in the subclavian vein. The ducts have valves to stop lymph from flowing down.

As the lymphatic system does not have its own heart, it does not experience any pressure. Therefore, lymph flows slower than blood. This gives the WBCs in it more time to look for germs and fight them.

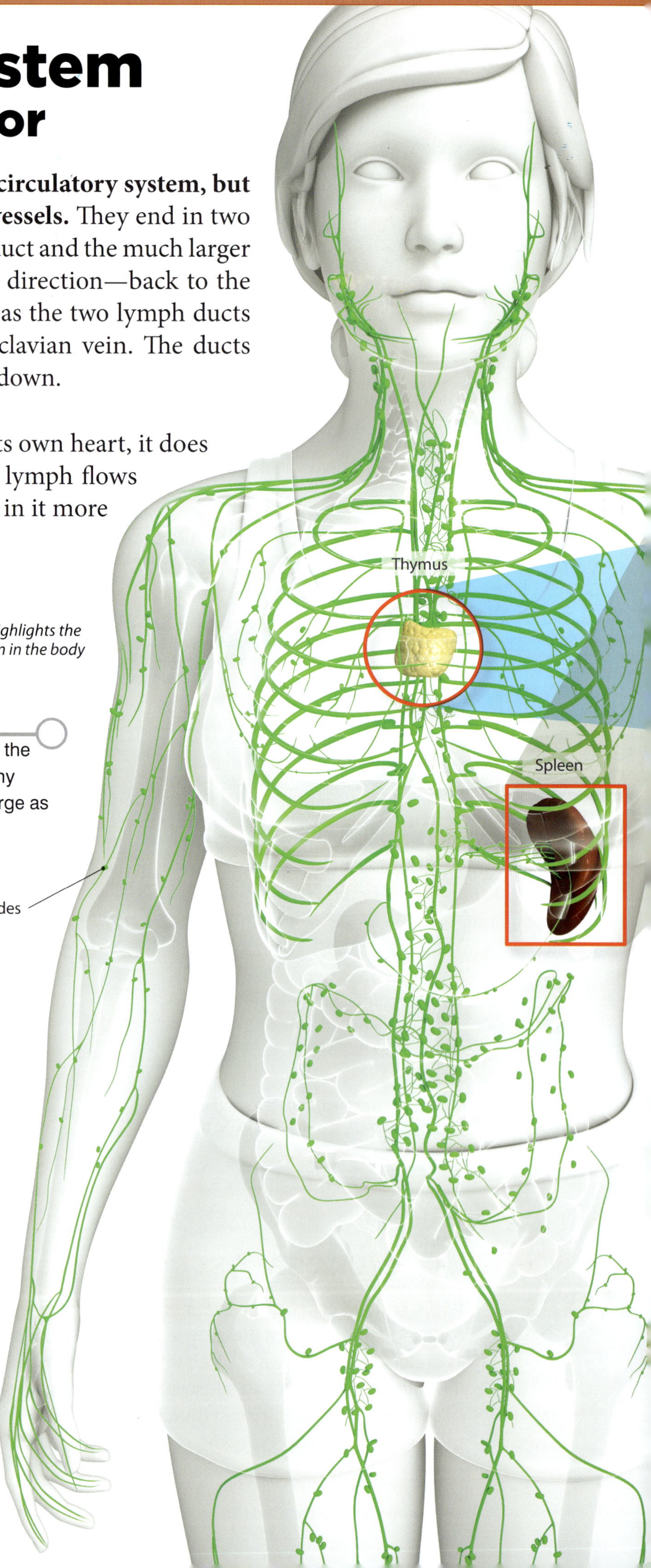

Thymus

Spleen

Lymph Nodes

The diagram highlights the thymus and spleen in the body

Lymph Nodes

Hundreds of lymph nodes do the work of pumping the lymph as it drains out of tissues. They come in many sizes: some as small as a pinhead, and others as large as kidney beans.

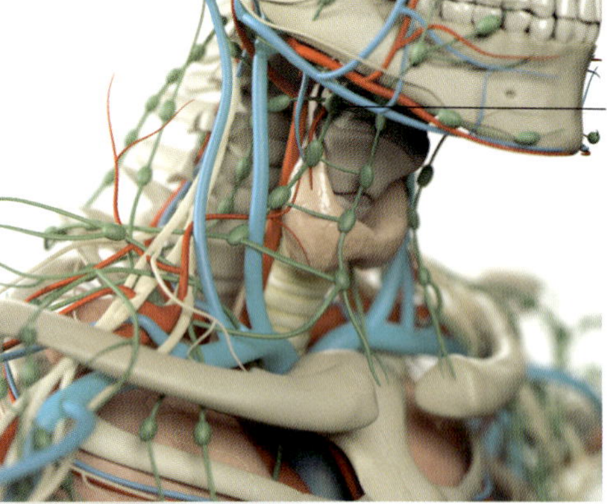

A 3D representation of how lymph nodes appear in the body

They also act as filters as lymph carries all the rubbish from the tissues. The filtered rubbish is attacked by WBCs, thousands of which sit in the lymph nodes. This also helps WBCs to learn and identify germs which may be trying to hide. Hence, the lymphatic system has more WBCs than the circulatory system, which is why they are also called lymphocytes. Some parts of your body have more lymph nodes than the others, like the armpits, neck, throat, and groin.

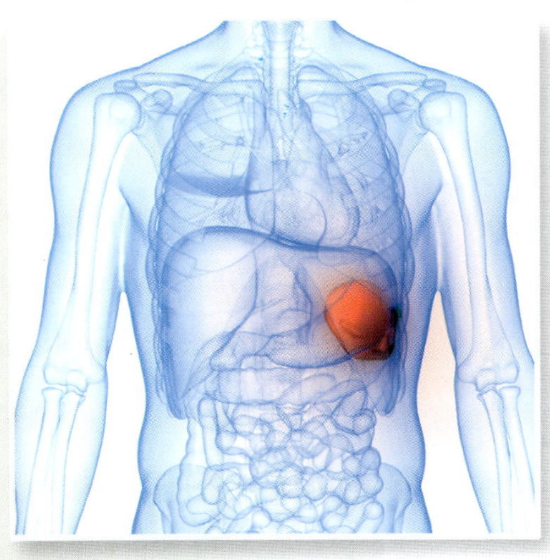

Spleen

Your spleen is a very important organ, which cleans your blood. You will find it on your left side, just behind and below the stomach. It looks like a very large lymph node, and it does many things a lymph node does. But it does something else too; it filters blood and removes old and worn out RBCs. It then destroys these and recovers the sugars, proteins, and lipids in them to be recycled by the body to make new cells.

The spleen also makes some WBCs and trains a whole lot more to detect germs. So it is a part of the immune system too.

The diagram highlights the spleen in the body

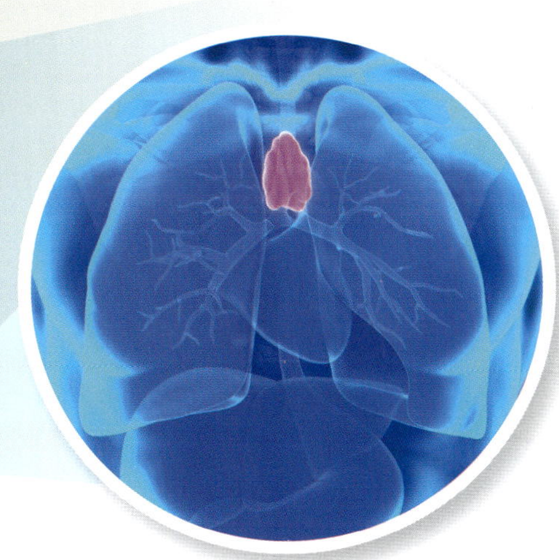

Thymus

The thymus is another organ of the lymphatic system. You find it between your breastbone and your heart. Like the spleen, it also filters blood and removes dead RBCs. But its main job is to make a kind of WBC called T-cells, which are born and trained here. T-cells are like commando cells. They do many things to fight germs especially as a part of the immune system, which protects our body.

The diagram highlights the thymus in the body

Isn't It Amazing!

The thymus is the only organ in your body that actually decreases in size as you grow older, which means T-cells stop training after childhood.

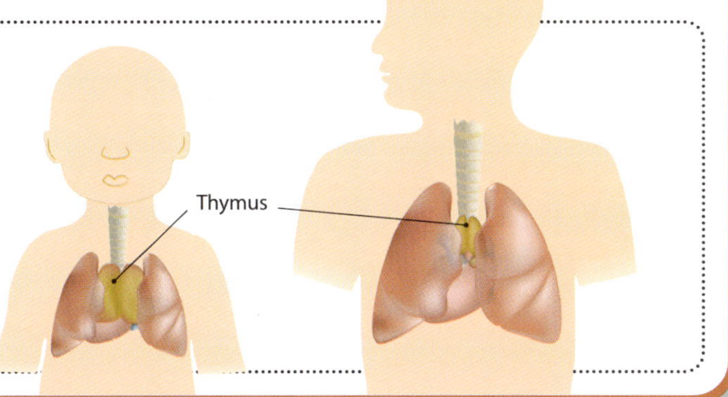

The thymus gets its name from its silhouette, that looks like a thyme leaf"

Mucosa-Associated Lymphatic Tissue (MALT)

Some parts of your body need more help from the immune system than the rest, because they are exposed to germs from the air, like your mouth, throat and lungs; or from food, like your intestines. Therefore, the body has evolved to put more lymph nodes and lymph ducts in these places, close to the lining of the lungs and intestines called the mucosa. This complicated tissue is called Mucosa-Associated Lymphatic Tissue (MALT). It is full of lymphocytes which are ready to fight any germs that try to enter the body.

Our Skin: The Great Wall

Did you know that the largest organ of the body is your skin? It is the first defender of the body. It keeps the internal organs safe from the ill effects of varying temperatures. Our skin is made up of three layers—while the outermost layer is called the **epidermis**, the inner layer found beneath the epidermis is known as **dermis**. It contains hair follicles, sweat glands, and connective tissue. The third layer is the **hypodermis**, a deeper subcutaneous tissue which is made up of fat and connective tissue. Your skin is also your body's first immune warning system. It can stretch and also squeeze itself. That is why your skin does not tear when you extend your arms during a stretch or crouch down to find a coin that has rolled under the bed. Your skin has nerves that tell you when something has touched it.

Keratinocytes

Keratinocytes are wonder cells found in the epidermis that perform many functions. These cells are tightly packed together to keep out foreign particles like bacteria and viruses. They make vitamin D in the presence of sunlight. Keratinocytes also make **keratin**, which is needed to form the hair and nails. They detect wounds and heal the skin. They alert the immune system to destroy particles that might have entered from an open wound.

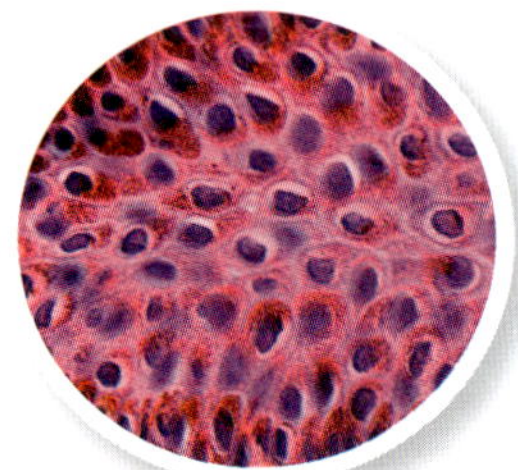

▲ *90 per cent of epidermal skin cells are Keratinocytes*

Hair Follicles

Hair follicles are the organs from where hair grows. Hair is made of fibres of a protein called keratin. The hair on our bodies is of two types. First is the long wiry hair which we can see and which keeps us warm by forming a carpet that traps air. When we are cold or scared, the hair stands straight, causing 'goose bumps'. It then traps a thicker carpet of hair.

Our skin also has shorter, white hair, which cannot be seen without a magnifying lens. These act like burglar alarms. If a louse, bedbug, fly, or mosquito brushes against them, they trigger the nervous system, which in turn makes us 'itch'. We instinctively scratch and scrape off the creature.

The Inner Skin

In the dermis, there are **sweat glands**, oil glands, hair follicles, and blood vessels. Sweat glands keep the body cool by releasing water along with some salt and amino acids onto the skin, which evaporate and cool it. However, if we sweat too much, a lot of amino acids collect on the skin. Bacteria can grow on these and, in turn, produce smelly chemicals. That is why when sweat collects in the armpits, it begins to smell bad.

The oil glands are also called **sebaceous glands**, because they produce an oil called **sebum**. It spreads all over our skin and makes sure that water does not enter skin from outside. That is why, even if you get wet in the rain, you do not soak up water like a sponge. Sebum also prevents bacteria and fungi from growing on the skin.

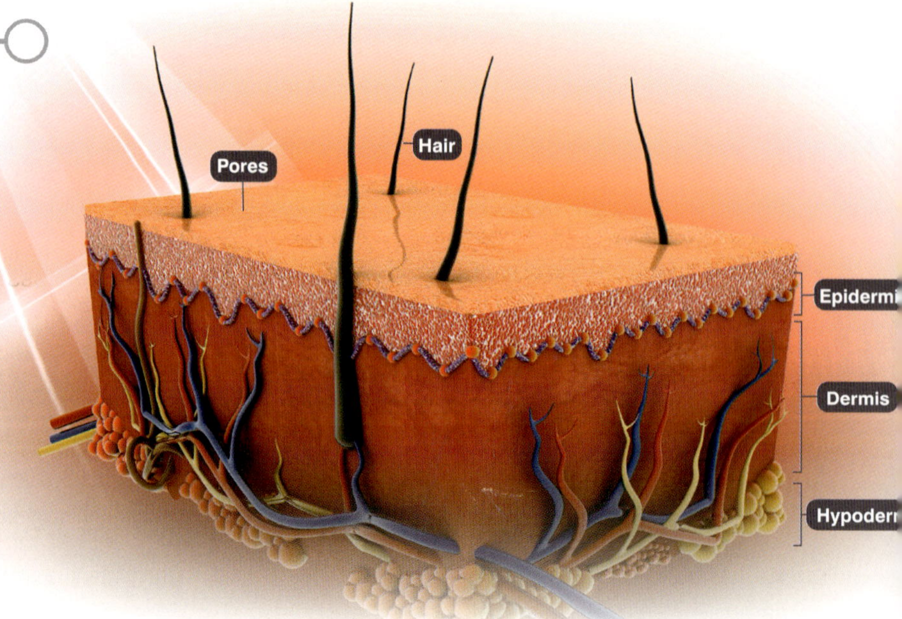

The Organs that Protect Us

The immune system performs the critical role of defending the body against foreign bodies, viruses, bacteria, etc. However, it may surprise you to know that the immune system is unlike the other organ systems that make up the human body. That is because it has no organs dedicated solely to itself, as is the case with the digestive system or nervous system. Instead, the organs of the immune system work towards safeguarding the body in addition to performing other functions.

Lymph Nodes

The **lymph nodes** are part of the **lymphatic system**, which is our body's second circulatory system. This system carries lymph, a colourless liquid that contains the waste products of our tissues. Lymph passes through a number of tiny organs called lymph nodes, which filter the lymph and remove **pathogens**, broken cells, etc. which are then destroyed by the immune system, similar to how the spleen works.

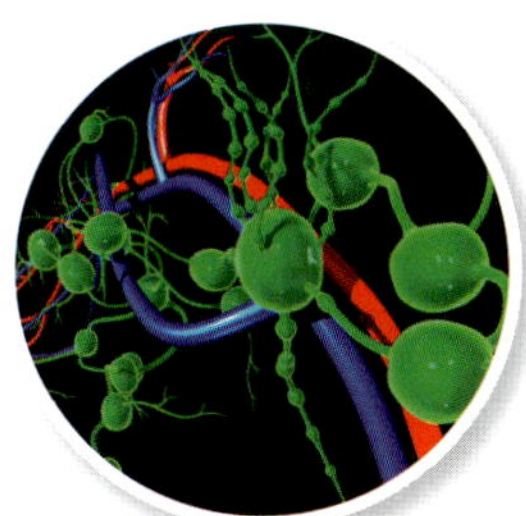

▲ *Lymph nodes filter germs and train WBCs*

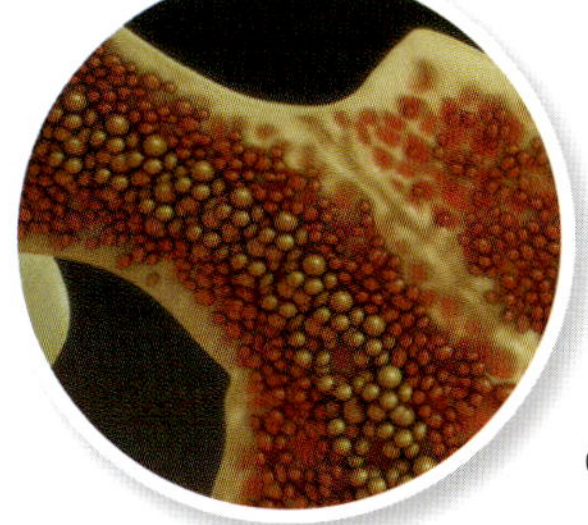

▲ *Bone marrow makes up 4 per cent of an adult human's weight*

Bone Marrow

This is the fleshy tissue inside most large bones. This is where red blood cells (RBCs), white blood cells (WBCs), and platelets are made. Their parent cells are called stem cells. The bone marrow also trains some of the WBCs to fight disease-causing germs called pathogens. These trained cells are called B-cells.

Tonsils and Adenoids

These are the lymph nodes that are present in our throat. They help to fight off infections in our mouth and nose respectively. Like all lymph nodes, they swell up during an infection. Unfortunately, this swelling can cause problems with swallowing food or breathing. That is why you have to have your tonsils removed if they swell up too much. The swelling of adenoids causes snoring, because they block the passage of air from the nose into the lungs.

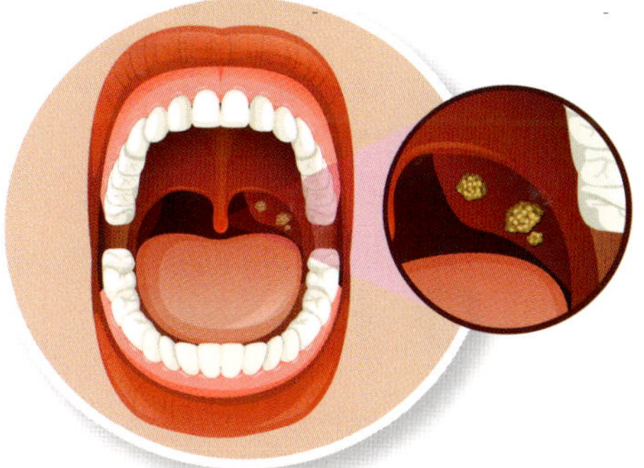

▲ *Your tonsils are located at the end of your mouth*

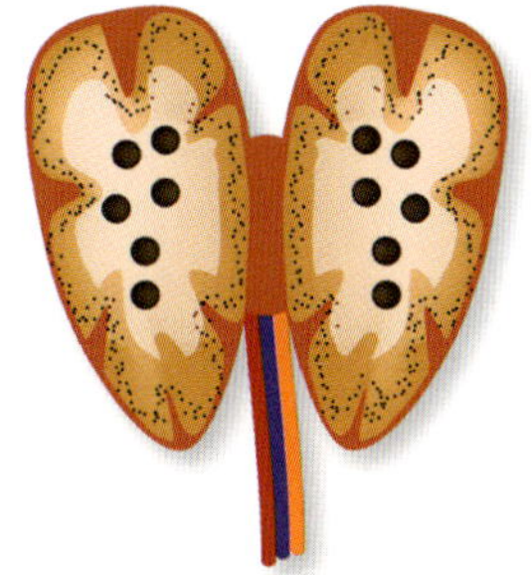

▲ *The thymus trains WBCs to recognise germs*

Thymus

The thymus is a tiny organ between the heart and the breastbone. It performs many tasks for the body, including training a class of WBCs called T-cells. The thymus also acts as a resting camp for other kinds of WBCs.

Spleen

The spleen is a tiny organ near your stomach. It is a resting place for platelets, though its main job is to break the worn-out RBCs. It acts like a filter where the good RBCs are separated from damaged RBCs, platelets, pathogens, and other stuff. The spleen needs phagocytes to attack RBCs and destroy them.

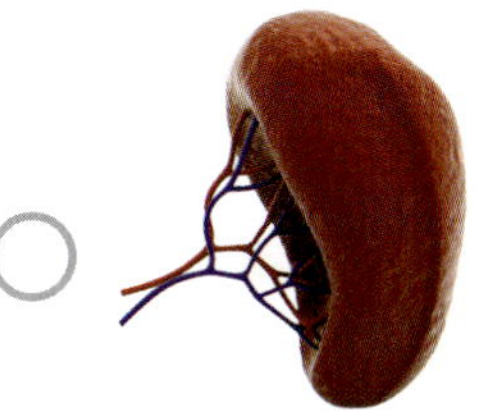

▲ *The spleen filters blood, kills dead RBCs and acts as a training ground for WBCs*

You are Under Attack

Our body can be attacked in two ways: either by pathogens or allergens. Pathogens are living creatures like viruses, bacteria, fungi, plasmodia, and worms. Allergens are often non-living things like dust and certain kinds of chemicals, as well as living things like pollen, nuts, vegetables, or meat products.

Pathogens

Many pathogens invade our body when our immune system is weak. Doctors call them facultative parasites. Some, like the one that causes malaria, however, must invade our bodies because they can only reproduce that way. These are called obligate parasites.

There are other microscopic creatures that are called commensals. They hang around on our skins or our intestines, living off sweat, and undigested food. Some of them help us fight off pathogens and make the vitamins we need.

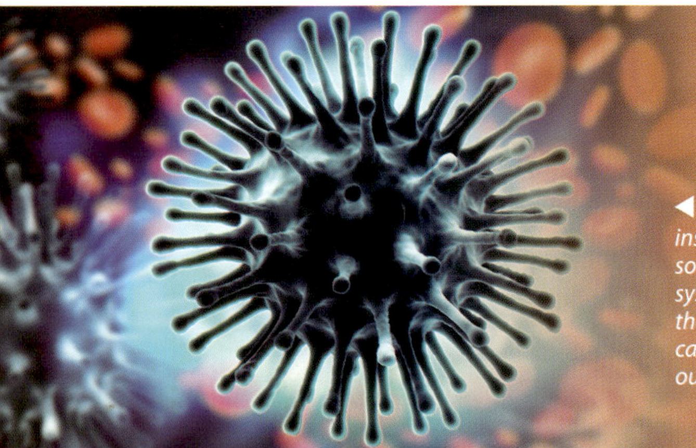

◀ *Viruses hide inside human cells, so the immune system must catch them before they can get inside our cells*

Viruses

Viruses are so tiny that you cannot see them with the naked eye. They are the tiniest living organisms. They are simple organisms made of DNA covered by a coat of protein—we are not even sure that they are really living things. Viruses are the cause of different kinds of diseases, from common cold to swine flu, dengue fever, and even HIV/AIDS.

Fungi

Most fungi are not infectious. But those that are, infect the tongue, causing diseases such as white-tongue or the skin, leading to infections like ringworm. Fungal infections often happen when the immune system is weak, especially when the body is recovering from a major illness.

▶ *Cells of the pathogenic fungus Candida stick to the tongue, making it look white*

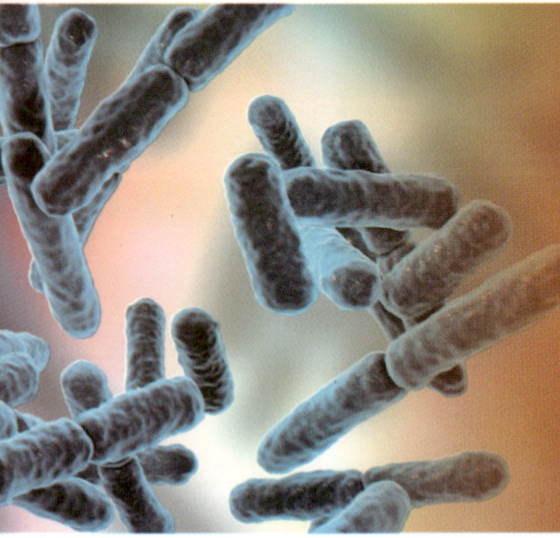

◀ *Bacteria come in many sizes and shapes, from round, pea-like cocci to rod-like bacilli, spirals, and threads*

Bacteria

Bacteria cause most of the diseases known to us, from pimples on the skin to fevers and cholera, leprosy, and tuberculosis. Bacteria are the simplest kinds of cells, with just cytoplasm covered by a cell wall.

There are two types of bacteria—those that need oxygen, called obligate aerobes and those that do not need oxygen, called obligate anaerobes. *Bacillus subtilis* is an obligate aerobe. *Clostridium* is a rod-shaped bacterium found in the intestinal tracts of animals and human beings. It does not need oxygen to grow.

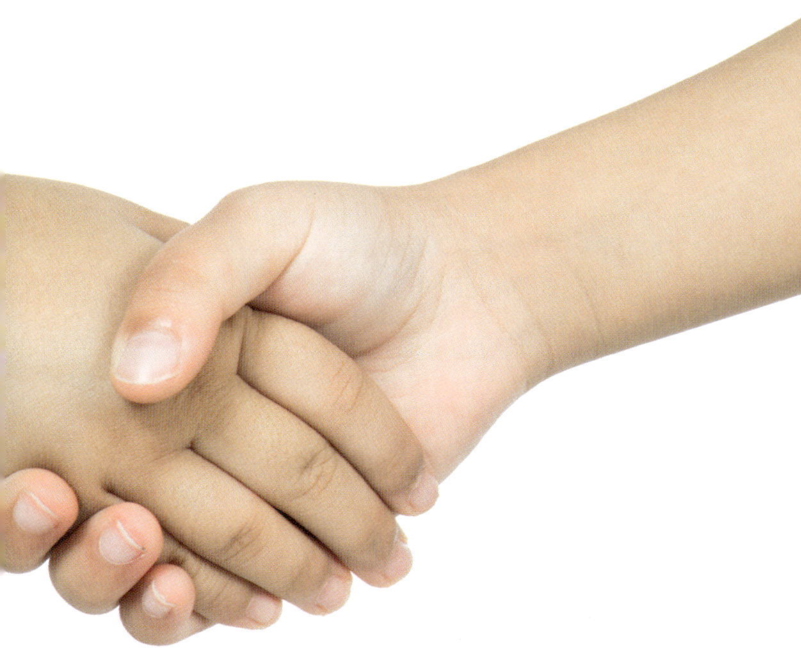

Isn't It Amazing!

Viruses need to get inside another living thing to make more viruses. Therefore, they will infect any living thing, from human beings and other animals to plants and one-celled creatures like amoebae and even bacteria. These last ones are called **bacteriophages**.

Scientists are using bacteriophages to treat some infections caused by bacteria

Worms

If you think of worms, you are probably thinking of earthworms. But the worms that attack human beings are really tiny, and are called **helminths**. The most common of these are tapeworms, which infect the intestine.

Some helminths drill through the intestine wall and get into the blood, through which they can reach the brain

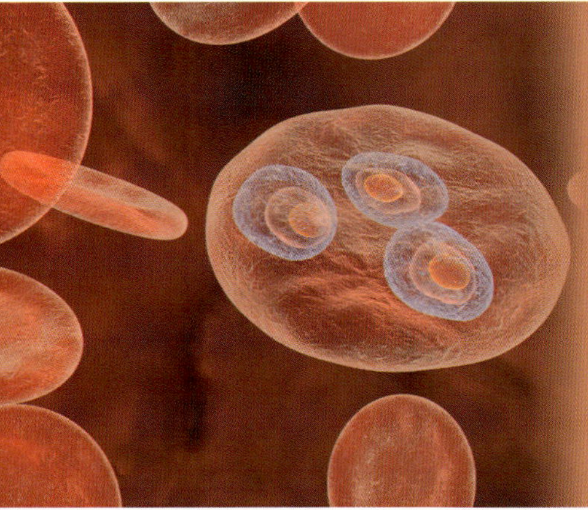

Plasmodium hides inside red blood cells to escape the immune system, so it is a hard enemy to fight

Protozoa

Protozoa are one-celled animals that have a nucleus and no cell wall. Protozoa cause some of the deadliest diseases known to us, like malaria, caused by Plasmodium, and sleeping sickness caused by Trypanosoma. Protozoa cannot infect us without an agent, which is often an insect that bites us, like the mosquito. The insects live in tropical regions, so the diseases they cause are called tropical diseases.

Pollen

While pollen do not really attack or invade our bodies, when they are in the air, they get into our systems through breath and food. Pollen triggers the allergic response of the immune system in some people, which causes hay fever, leading to sneezing, reddening of the eyes, and itching in the throat.

Allergic patients need to be careful in spring, as repeated exposure to pollen can cause death

Our Immune System's Chemical Toolkit

Modern armies have all sorts of weapons and gadgets to fight a war—including radars and satellite phones, night vision goggles, and jammers to block enemy communications. Our immune system also has gadgets like these, except that they are all chemical in nature. These tools give the immune system an incredible ability to act against any infection. They help the immune system increase or decrease the strength with which it responds, and also helps coordinate among all its cells. This is known as adaptability. They also prepare the rest of the body for fevers, and warn cells to look out for pathogens. During organ transplants, doctors give patients drugs called immuno-suppressants. These make sure that these immune chemicals do not attack the new organ.

Hormones

Many immune cells work in the same way as hormones in the body. Hormones are chemical messengers that travel through the blood and bind to proteins called receptors on the surface of the cells.

If a receptor is a lock, a hormone is like the key that opens it. The same key can open different locks, i.e. the same hormone can interact with different kinds of receptors on different cells. Some hormones signal immune cells to stop being active, while other hormones make inactive immune cells active.

▲ *Model of serotonin, the key hormone that stabilizes our mood and feelings of well-being*

Granzymes & Perforins

These are enzymes that are stored in the granules of natural killer cells and also killer T-cells. When released, the perforins drill holes in the membranes of infected cells, causing them to burst. The granzymes (enzymes from the granules) then digest all the material.

▶ *Perforin, a protein responsible for pore formation in the membranes of target cells*

Cytokines & Interleukins

These are the messenger proteins of the immune system. There are many kinds of cytokines, each of which carries a special message. Cytokines can be made by most cells of the body, but only when infected. They are also made by macrophages that have found antigens. Both are received by T-cells and B-cells that then get activated to do their jobs.

Interleukins are cytokines that are used by B-cells and T-cells to talk to each other. They are mostly made by helper T-cells.

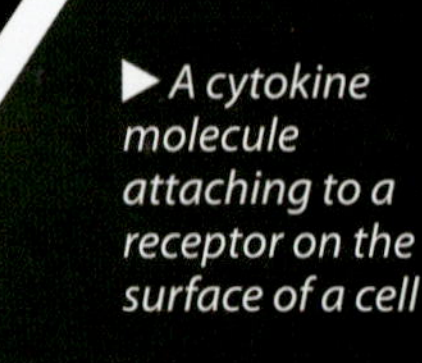

▶ *A cytokine molecule attaching to a receptor on the surface of a cell*

Interferons

These are messenger proteins that are released by the cells infected by viruses. They communicate to the immune system to send killer cells and macrophages to finish them off.

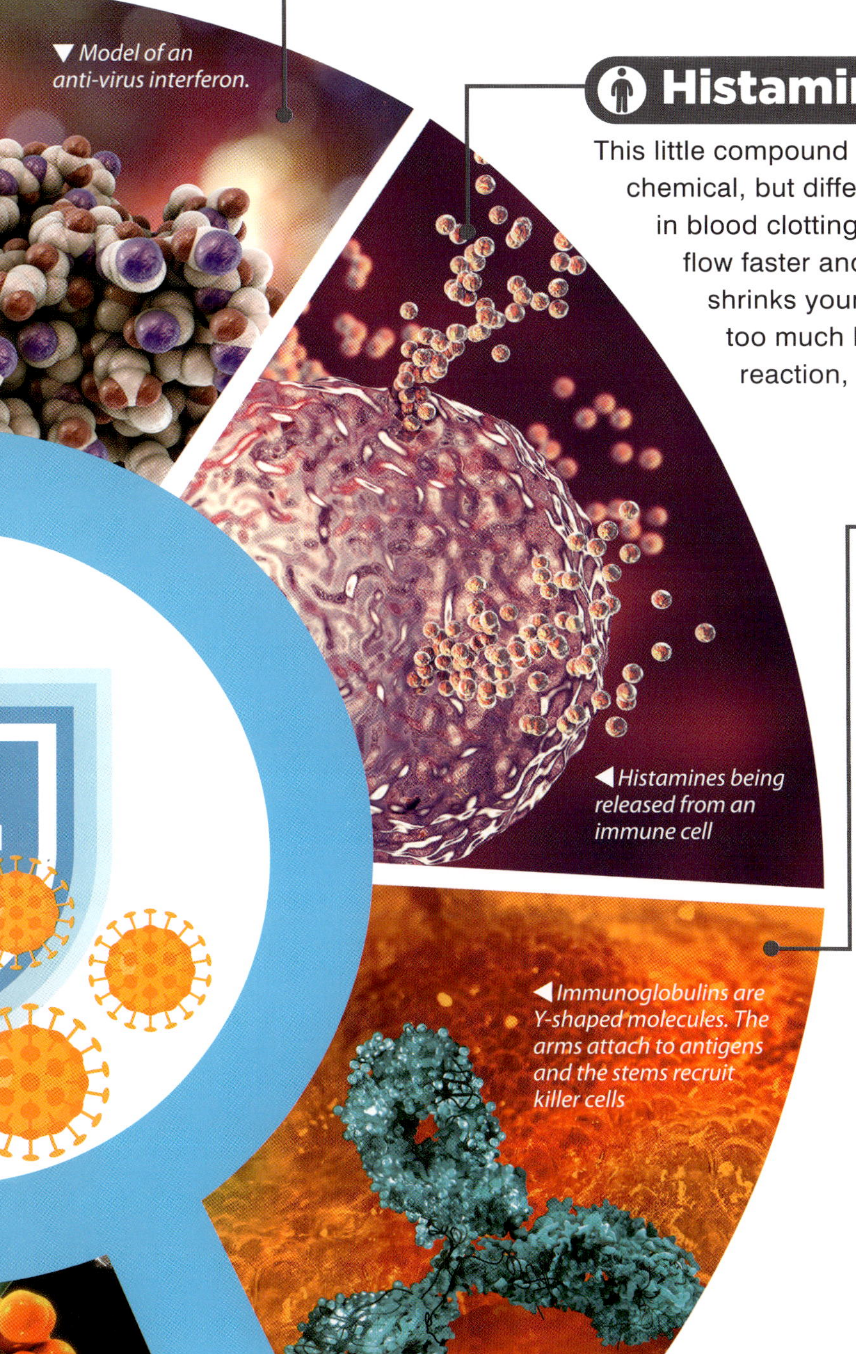

▼ Model of an anti-virus interferon.

◀ Histamines being released from an immune cell

◀ Immunoglobulins are Y-shaped molecules. The arms attach to antigens and the stems recruit killer cells

Histamine

This little compound is the body's alarm siren. It is a messenger chemical, but different cells read it differently. Histamine is involved in blood clotting, increasing heartbeat and making the blood flow faster and widening blood vessels among other things. It shrinks your lungs so you do not breathe in germs. However, too much histamine in the body causes an excessive allergic reaction, called anaphylactic shock.

Antibodies

These are the body's wonder weapons, and are made by activated B-cells. The medical term for them is **immunoglobulin**, Ig for short. They can be of the following five types:

IgM: This is the antibody made by B-cells the first time there is an infection.

IgG: These antibodies are made the second time there is an infection. IgG is made in huge quantities and can finish off the pathogen very quickly.

IgA: This antibody is released into the mucus of the intestines, lungs, and tears to fight pathogens before they enter the body.

IgE: This antibody is meant for allergic reactions, though it works like the rest.

IgD: This antibody stays on the surface of the cell to help it recognise the pathogen.

In Real Life

Babies don't just get nutrition from their mother's milk but also immunity. Immunoglobulin A travels into the baby's stomach, where it protects the baby from many diseases. Doctors call this passive immunity.

▲ Feeding on mother's milk is important for immunity among all mammals

Vaccines: A Jab of Safety

When the vaccine enters the body, macrophages catch hold of it, and eat it up by phagocytosis. Some of the bits of the dead germ are shown on the outside of the cell, and T-cells and B-cells try to recognise the antigen. Those who do, become activated. Most will fight the vaccine, but a few will turn into memory cells. So, your body is tricked into fighting a disease that wasn't there, but it left you vaccinated.

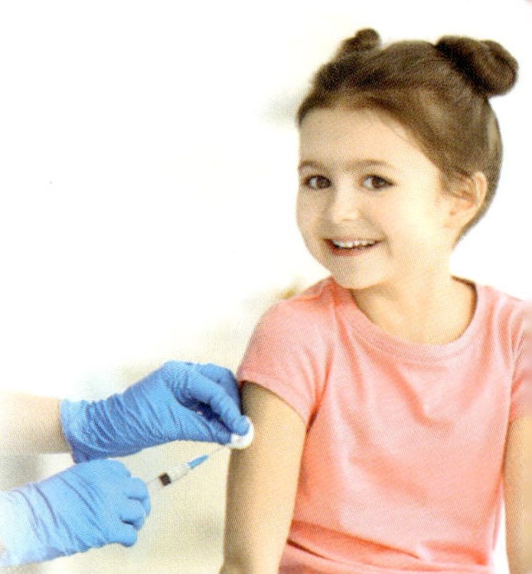

History of Vaccines

Edward Jenner (1749–1823) was the first to discover that having one disease can save you from others. He saw that the people in his village who got a disease called cowpox, did not get the much deadlier smallpox. Cowpox leaves little sores on the skin, and if you scrape material from these sores and inject them into healthy people, they too become immune. His discovery became a revolution in Europe; in 1811, Emperor Napoleon of France had his entire army vaccinated!

After Jenner, Louis Pasteur in France and Robert Koch in Germany continued to do a lot of research and developed a large number of vaccines against rabies, anthrax, cholera, and many other diseases. Most vaccines have to be injected, but polio is given as drops in the mouth. You get polio by drinking contaminated water, and your body fights it in the intestine's MALT.

Today, the government and doctors recommend that we give several vaccines to our children in order to help them become immune to many diseases.

▲ *A child getting a polio vaccine in Brazil*

Cancer Vaccines

Today, researchers have found that some kinds of cancers can be prevented by making antibodies against them. It means we can make vaccines against such cancers. For now, we have vaccines against cervical cancer, but many more are being tested.

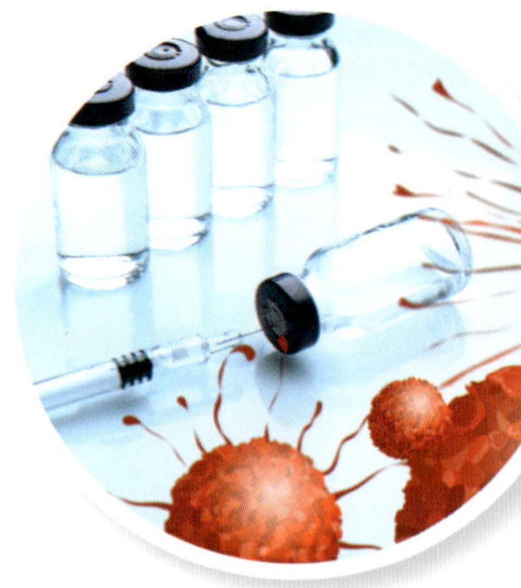

▶ *Vaccines for many viral diseases are still being developed*

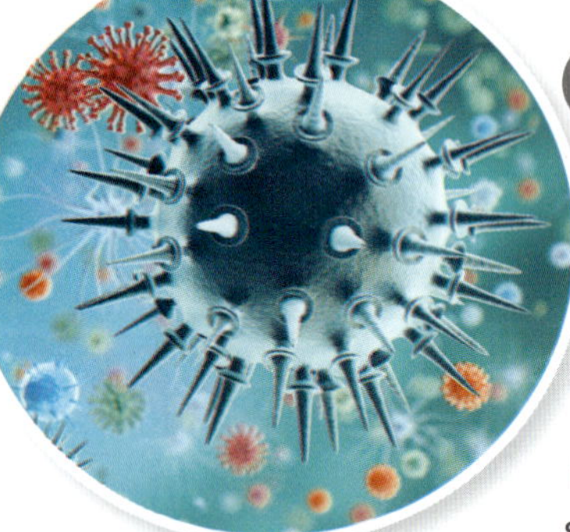

▲ *This is what an influenza virus looks like*

Flu & HIV Vaccines

Influenza is a disease that is caused by many kinds of viruses. Because of this, we don't have a single vaccine that works against it. Also, many influenza viruses mutate, and their antigens keep changing, baffling the immune system. The HIV virus also does the same thing.

But there are different vaccines coming up against flu, and doctors say that you should get a jab every year to stay safe.

Helping Our Immune System

Healthy food keeps the body fit, and generally helps the immune system. Fighting illness needs a lot of energy. You also need a lot of vitamins and minerals to ensure that the immune system is healthy and is able to make enough of its messenger molecules and chemical weapons. But most of all, you need to exercise a lot to keep your heart and circulation system healthy, so there's enough oxygen for all cells of your body.

Eating Healthy

What the immune system needs most are Vitamins K and D. Vitamin K is a necessary part of the clotting process. Without it, blood would not clot, and you could bleed uncontrollably. Good sources of vitamin K are green leafy vegetables, cheese, and eggs.

Vitamin D is necessary for T-cells to function correctly. Vitamin D is made in our skin in the sunlight, but you can also get it from fish and mushrooms.

Eating vegetables rich in vitamins is as necessary for your body as exercising

Sleep and Exercise

Lack of sleep or a lot of stress in the body causes the production of steroid hormones, which reduce immunity. On the other hand, exercise boosts immunity. It keeps up healthy circulation of blood so, WBCs can travel faster and also reduces stress hormones. So, go out and play a lot!

Young children should be encouraged to play outside so that they can make friends and get some exercise

Hormones

Some steroid hormones in our bodies, like testosterone and cortisol, suppress the immune system. The body makes more of them when the immune system has fought off the pathogen, and energy is needed for other things. In autoimmune disease, these hormones are used to protect the body from being attacked by its own immune cells.

In Real Life

Did you know that listening to music for 50 minutes every day can help boost your immunity? This is because music helps calm the mind and reduces stress in the body. So, when you're studying or travelling, listening to music is a good idea!

Listening to music is good for reducing stress

Nose: Your Breath's Gatekeeper

Most of the respiratory system is made up of the vessels that bring fresh air right into the body. First there is the **conducting zone** which includes the nose, throat, windpipe, and air passages inside the lungs. The **respiratory zone** consists of the tissue where oxygen filters from the lungs into the blood.

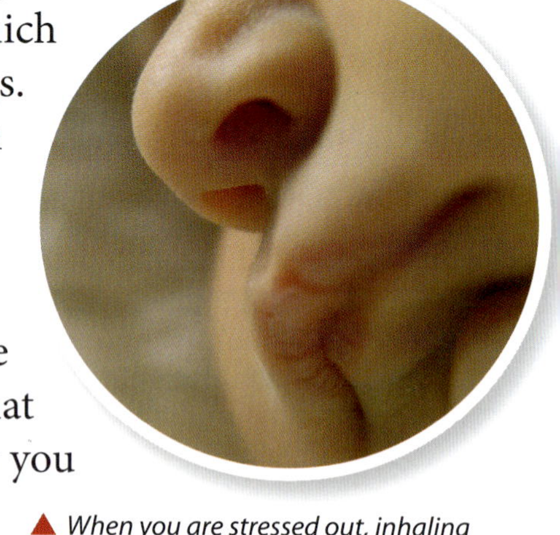

▲ *When you are stressed out, inhaling and exhaling makes you feel better*

The nose is the gateway of the conducting zone. There is more to it than the part that sticks out of your face, as it occupies a large space inside your head, between the brain and the mouth. It cleans the air that comes in through the nostrils, removing dust and germs. That is why you normally breathe through your nose, else air enters the lungs from the mouth.

Outer Nose

The part of the nose that you can touch and see is called the bridge of the nose. It is made of a cartilage septum and thickened skin that make the nostrils on either side. The inside of your nostrils is like the outer skin with sebum glands and hair. This hair keeps big dust particles, dirt, and even insects from entering the nose. Interestingly, human beings mostly breathe through one nostril at a time, even though we have two nostrils. If you want to check this fact, put your hands close to your nose and breathe in, then breathe out.

In Real Life

Some germs, like the flu virus, can infect your sinus. Your body's immune response causes them to swell. You can feel them on your face. This swelling causes facial pain, headache, and sometimes fever. It often follows a cold, leaving you with a runny nose.

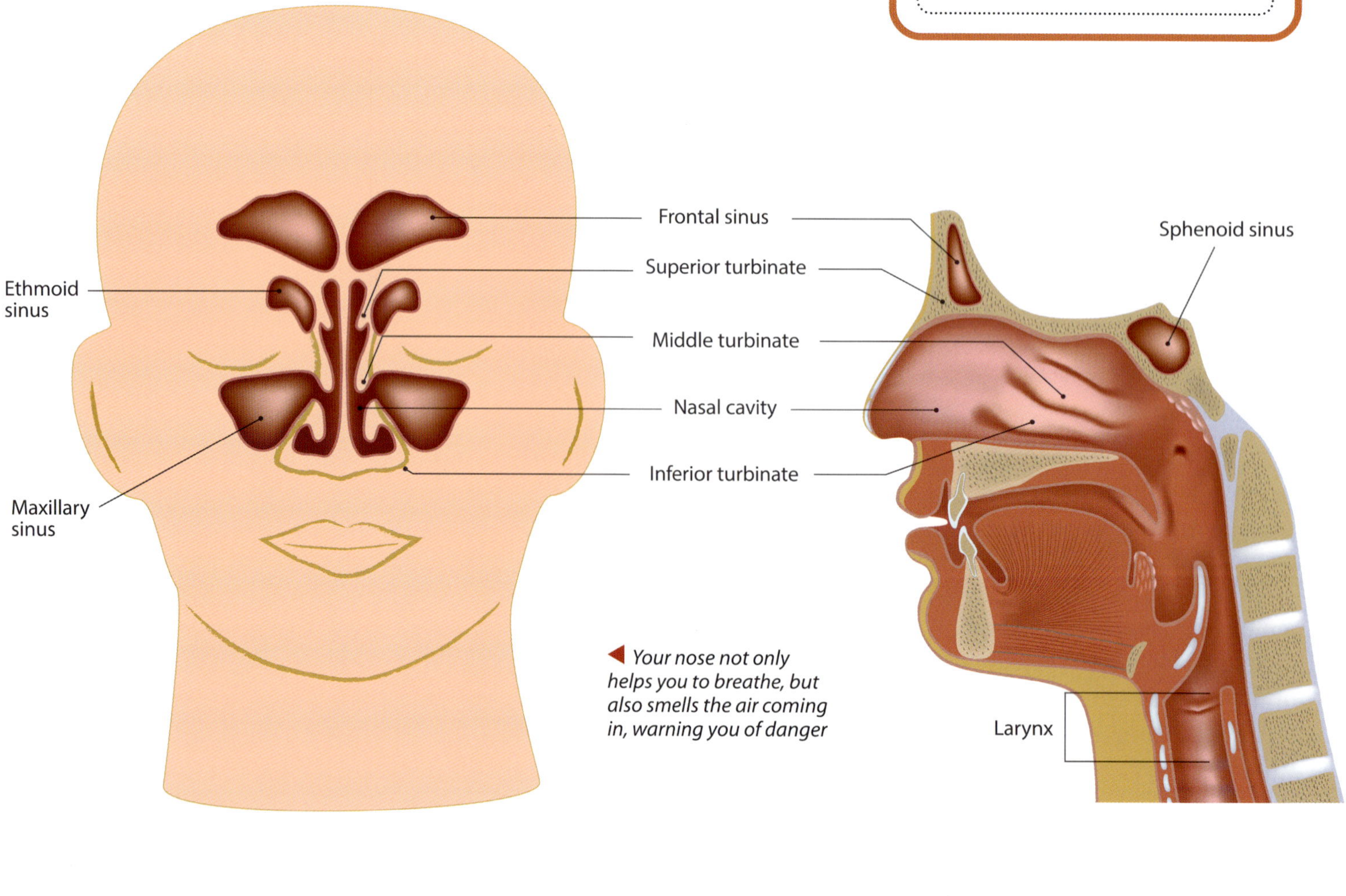

Frontal sinus

Superior turbinate

Ethmoid sinus

Middle turbinate

Nasal cavity

Inferior turbinate

Maxillary sinus

Sphenoid sinus

Larynx

◀ *Your nose not only helps you to breathe, but also smells the air coming in, warning you of danger*

Inside the Nose

The septum extends right through the inner nose, where it is made of bone. It is not always in the centre, but often slightly shifted to the right or left. The bony walls of the nose are folded into **conchae** or turbinates. They swirl the air within the nose, like turbines in a generator. When you are breathing out, the nasal conchae trap water vapour so your nose does not become dry. The inner nose also has extensions called **paranasal sinuses**, of which there are four pairs—the frontal, maxillary, sphenoidal, and ethmoidal. The air in them makes the skull light.

Incredible Individuals

Cleopatra, the Queen of Egypt, had a long nose. Blaise Pascal once famously remarked, that had Cleopatra's nose been shorter, the whole face of the world would have been changed. He was referring rhetorically to the collapse of the Roman republic , as both, Julius Caesar and Marc Antony were spellbound by her beauty.

▲ *A bust of Cleopatra*

Sneezing

Sneezing is our body's way of keeping itself healthy by expelling germs, pollen, and other foreign objects with great force. The hair in your nostril are sensitive to the smallest things and trigger the nervous system to make the muscles of the chest and throat contract very fast. When you sneeze, air is forced out of your mouth and nose together. As the air moves out, it carries mucus droplets with it, in which the germs are trapped. If somebody happens to be near you when you sneeze, they might inhale your germs. That is why you should cover your nose and mouth while sneezing.

Cleaning the Air

The nose works like your body's own air purifier. The nasal cavity and sinuses are lined with a special **respiratory epithelium**. This is made of cells that have microscopic hair called cilia. Scattered in between are **goblet cells** that make mucus. As the air you breathe in swirls, the cilia and mucus catch all the dust and germs in it. Under the epithelium is a special tissue called **nasal-associated lymphoid tissue** (NALT), which has immune cells in it. These destroy the germs caught by the cilia. The mucus makes the air humid, while blood capillaries under the epithelium warm the air before it goes into the lungs. The 'purified air' is now ready for the lungs.

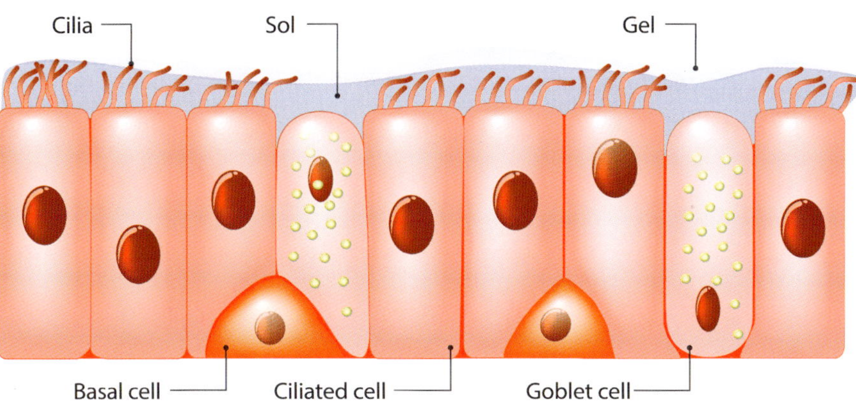

▲ *The nasal mucus is made of a sticky, outer gel that traps germs and a fluid inner sol that moves it*

Isn't It Amazing!

Birds do not have external noses. Instead, they have a pair of openings called nares just above their beaks that open directly into their inner noses. The nares can be used to tell male from female. For instance, budgerigar males have blue ones, while females have brown ones.

▲ *Some species' nares are covered with feathers*

Taking Air to the Lungs

Once air is cleaned and conditioned in the nose, it passes into the pharynx. This acts like a railway junction, where two tracks meet. The digestive system brings food from the mouth to the food pipe. The respiratory system brings air to the windpipe. So why do they meet at all?

You should avoid talking while eating because the food might accidentally enter your trachea and interrupt your breathing

Parts of the Pharynx

The pharynx is made of muscle that helps you inhale air and also swallow food. It is made of three parts—the **nasopharynx**, the **oropharynx**, and the **laryngopharynx**. The nasopharynx is the extension of the inner nose to the throat. It has the **pharyngeal tonsils**, also called the adenoids, which hang down from the top like a fold of tissue. It has another flap of tissue called the **uvula** that extends from the palate, which separates the nose from the mouth. While swallowing, the uvula stops food from entering the nose. The nasopharynx also has two canals called eustachian tubes that connect it to the ears so that they can be of the same pressure as the atmosphere, or else the ear drums will burst.

The oropharynx is the middle part, where the two tracks cross. The hyoid bone and the spinal column surround it. The hyoid attaches two sets of muscles. The first is the tongue inside the mouth. Under the tongue are the **lingual tonsils**. The other set are the muscles that control the **epiglottis**. The epiglottis shuts off the wind pipe while you swallow food. At the end of the palate, are the **palatine tonsils**. The final part is the laryngopharynx which leads the air into the trachea.

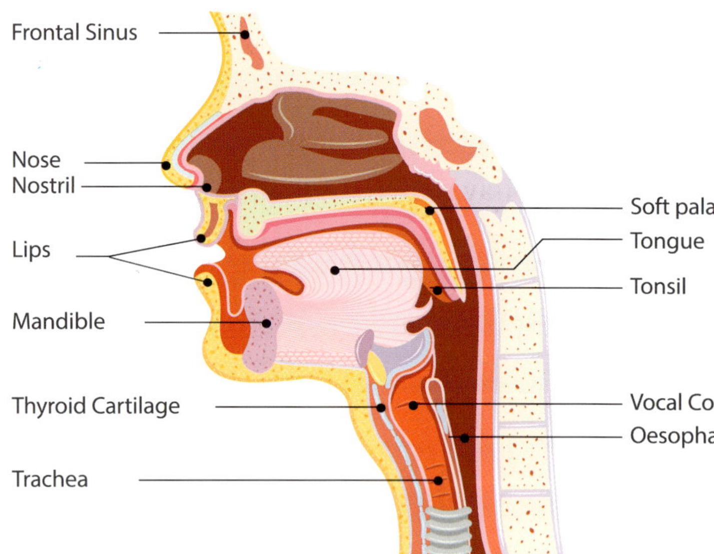

Tonsils are large in children but shrink as they grow older

Trachea

The trachea or wind pipe takes air from the pharynx to the lungs, sharing space in the neck with the oesophagus or food pipe. It is ringed by 20 C-shaped pieces of cartilage that make sure that it does not collapse onto itself like the oesophagus does. The **trachealis muscle** attaches to these pieces and, along with other tissues, makes the main air tube. It helps to stretch the trachea when breathing out, and shrink it when breathing in. In the lungs, the trachea branches into **bronchi**.

In Real Life

Infections acquired during childhood may leave you with enlarged tonsils. These cause obstructions to breathing that might need to be removed by surgery. If not, they can cause snoring or breathlessness.

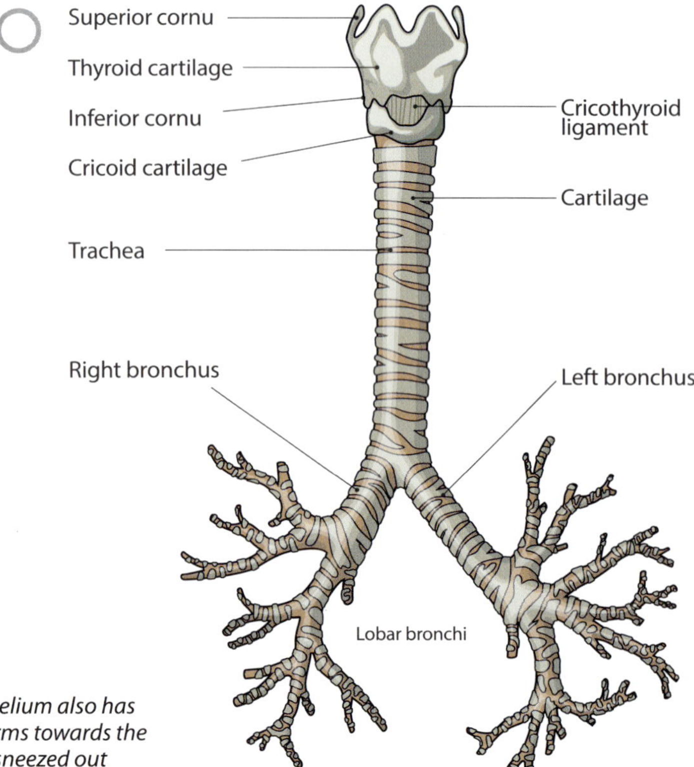

The tracheal epithelium also has cilia, which brush germs towards the nose, so they can be sneezed out

The Last Mile: Bronchi and Bronchioles

When the trachea reaches a point in the chest just below the neck, called carina, it splits into two bronchi (singular: bronchus) to the left and right. The carina has nerves that can sense whether anything other than air has fallen down the trachea. If there is, it will trigger coughing. Muscles in the trachea will try to push the body out before it enters the lungs. Otherwise, it may cause you to faint or obstruct your breathing.

▲ The bronchi and bronchioles make the bronchial tree

The bronchi enter the lungs, where they further split into secondary and tertiary branches. All of these have cartilage rings to stop them from collapsing. Doctors call the point of entry the **hilum**, where the nerves, pulmonary arteries and lymph vessels enter and the pulmonary vein leaves.

Isn't It Amazing!

Insects have a very different respiratory system. They have tiny pores under the bodies called spiracles, which let air into large sacs. Tubes called trachea take air around the body, branching into tiny tracheoles, which deliver oxygen to cells.

▶ Unlike vertebrates, an insect's respiratory and circulatory systems do not interact

Bronchioles

Bronchioles branch off from tertiary bronchi and further branch into terminal bronchioles. These feed the units of gas exchange in the lungs called **alveoli**. They have no cartilage.

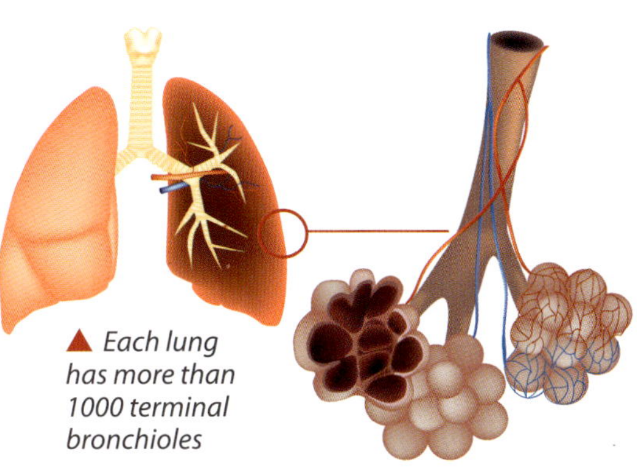

▲ Each lung has more than 1000 terminal bronchioles

Cross-section

The bronchioles are lined inside by epithelial tissue which has cilia to brush germs and dust out of the lungs, and also goblet cells that make mucus to help them. Surrounding it is a piece of loose tissue called the **lamina propria**. Like the nose has NALT, the lamina propria has BALT or **bronchial-associated lymphatic tissue**. It has immune cells in it, which destroy germs. This is surrounded by smooth muscle that keeps the air moving.

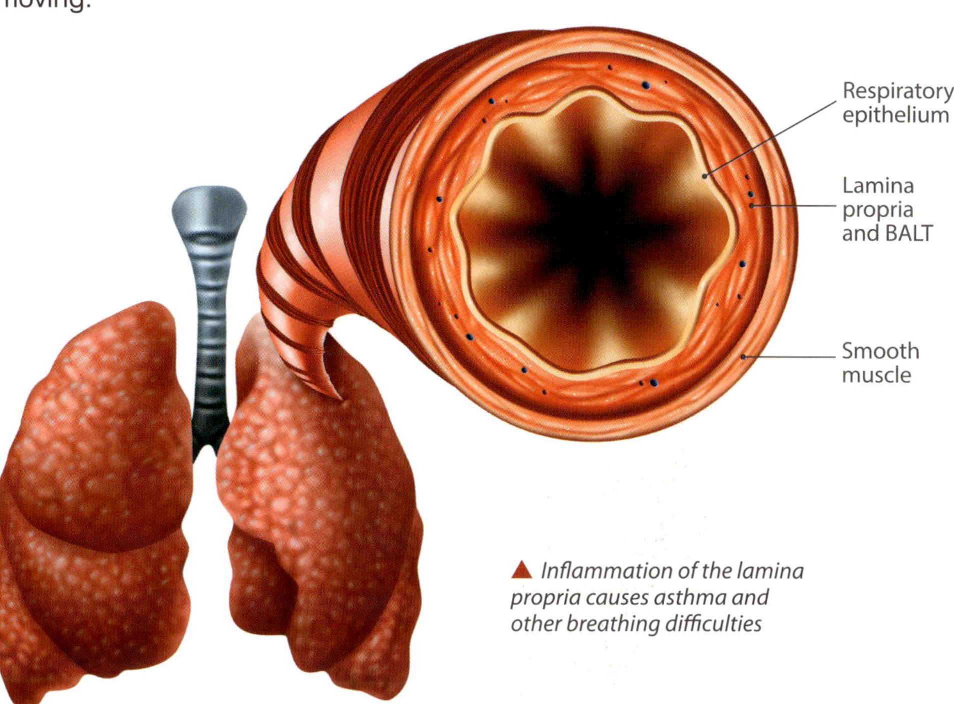

▲ Inflammation of the lamina propria causes asthma and other breathing difficulties

In Real Life

Infection in the bronchioles leads to a build-up of mucus, which causes congested lungs. This is called **bronchitis**.

Destination: Lungs

The air you take in from your nose has to reach the respiratory zone. Here, the respiratory system meets the circulation system. This zone has two parts, the **respiratory bronchioles** and the **pulmonary lobes**. It is made up of special tissue which allows oxygen to diffuse right into blood, where haemoglobin in the red blood cells is waiting for it. The blood has let go of carbon dioxide from deep within your body's tissues. It is now ready to make the journey to the outside world. Thus, you complete one breath.

Lung Structure

Your lungs come in a pair, with the left lung slightly smaller than the right one. Both lungs are encased in the ribcage and covered by a bag called the **pleura** (plural: pleurae), which cushions them against the ribs. The **diaphragm**, a giant, dome-shaped muscle, seals off the ribcage from below. The left lung has the cardiac notch, a space it makes for the heart to fit in.

Each lung is made up of lobes, which are served by secondary bronchi. The larger, right lung has three lobes, while the left lung has just two. Each lobe is divided into segments, which are then divided into lobules, each of which gets a bronchiole. Lobules are divided into alveolar sacs that, in turn, have alveoli within them. The alveoli are the functional units of the lungs.

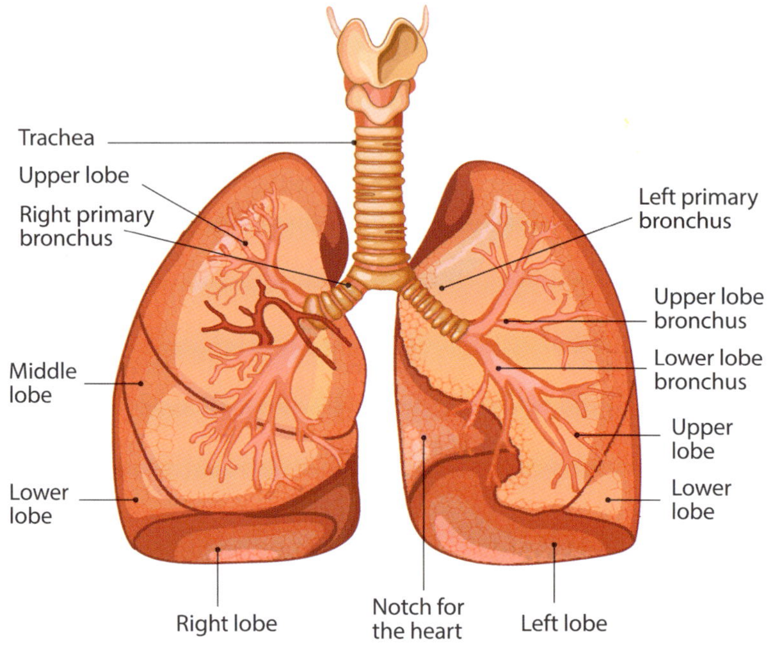

▲ *The right lung is broader than the left lung*

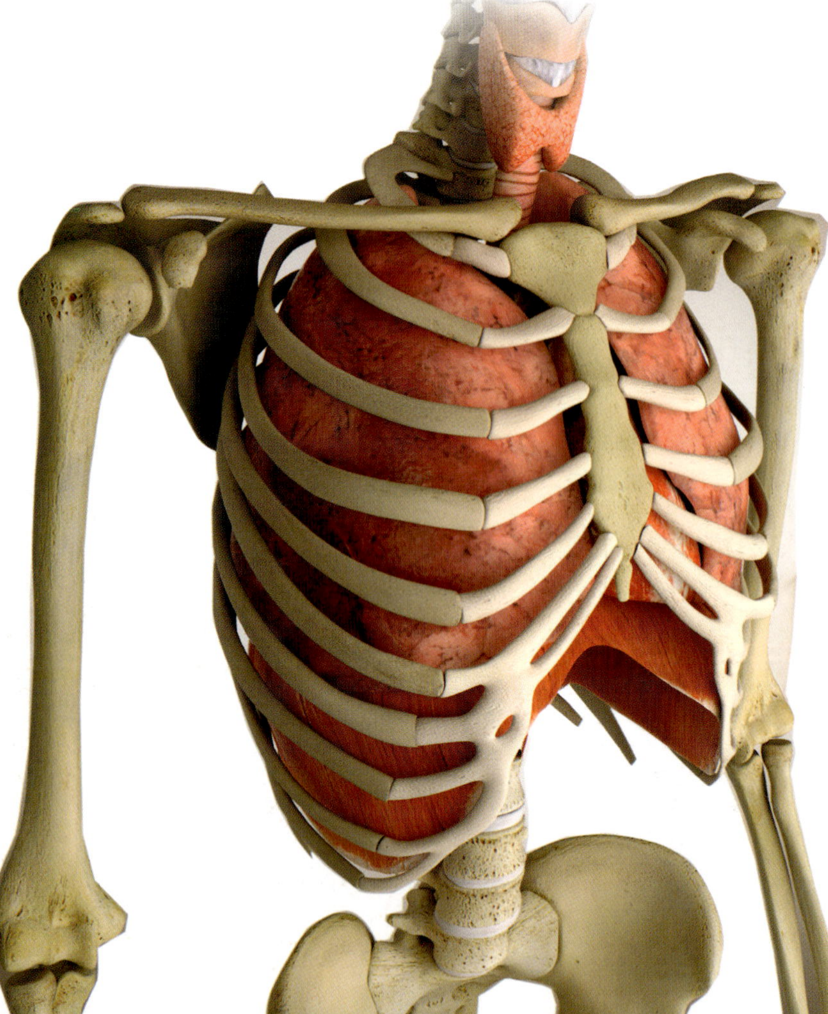

Pleurae

The pleurae that cover your lungs are made of two layers. The inner layer is visceral pleura, that lines your lungs directly, while the outer layer, that is, parietal pleura, links them to the rib cage, sternum, and diaphragm. Between them is the pleural cavity, which is full of air. Pleurae act like a cushion, when the inflated lungs push against the rib cage. They also protect the lungs against infections. The pleurae may swell up due to infection, leading to a condition called pleurisy. It may cause dry cough and difficulty in breathing. Sometimes, the pleural cavity may be filled with too much air, pushing pressure on the lungs. This is called pneumothorax. Injury to the lungs may cause them to be filled with blood, leading to a condition called haemothorax.

◀ *The pleural sac cushions the lungs while breathing*

Inside Your Lungs

In many invertebrate organisms, the respiratory system does not interact with the circulatory system. It does so in molluscs, where **haemocyanin** molecules in the haemolymph receive oxygen from the gills. Red blood cells are present in fish to carry oxygenated blood from the gills to the heart, then through the body, and bring back deoxygenated blood to the gills again. In reptiles, birds, and mammals, the lungs take over and there are separate arteries for oxygenated blood and veins for deoxygenated blood. This brings a lot more oxygen to the tissues, so they can have a lot more energy.

Blood Vessels and Nerves

Lungs are the only organs that receive deoxygenated blood from the heart's **pulmonary arteries**, which come straight to them from the heart's right ventricle. The artery enters at the hilum and follows the **bronchial tree**, finally branching into arterioles. At the alveoli, the blood flows in tiny capillaries, ready to receive oxygen. They collect into venules, which follow the bronchial tree till the hilum, where the **pulmonary vein** leaves to enter the heart's left atrium.

The lungs are controlled by the parasympathetic and sympathetic nervous systems. The parasympathetic, also known as the 'rest and digest' system constricts the bronchial tree, causing deeper breathing, like when you are sleeping. The sympathetic or 'fight or flight' nervous system dilates the bronchial tree, so you have more oxygenation.

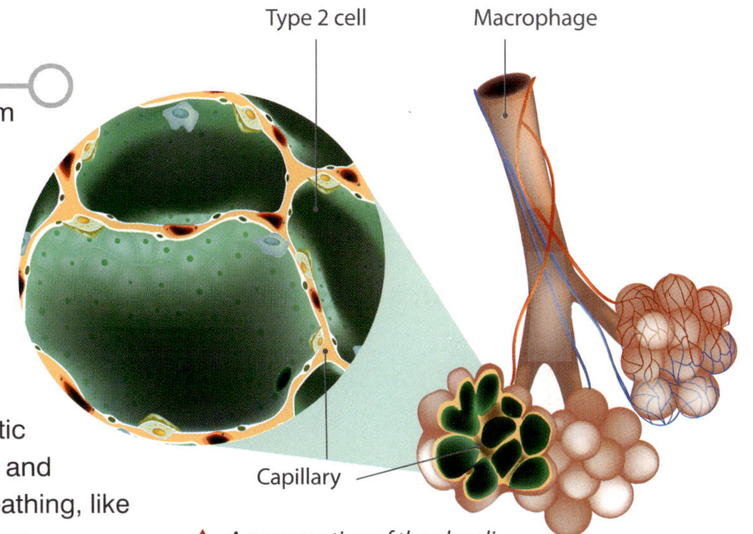

▲ *A cross-section of the alveoli showing walls and capillaries*

Alveoli

These are the working parts of the lungs and look like a bunch of grapes under a microscope. Each alveolus is a sack of respiratory cells, just 0.2 mm thick, with walls that stretch when full of air. They link to each other by alveolar pores, which make the air pressure in them equal. Most of the wall is made of type I alveolar cells, which let oxygen in and carbon dioxide out. A few cells belong to the type II alveolar cell group, which make **pulmonary surfactant**. This is a soap-like substance that makes it easier for oxygen to dissolve in the water of plasma. Immune cells called **alveolar macrophages** roam around the alveoli, snapping up any germs that manage to get there.

Alveolar Pressure

During quiet breathing, the air pressure in your alveoli, that is, the **alveolar pressure** will match atmospheric pressure, which is 760 mm Hg at sea level. However, it changes with the phase of breathing. When you are inhaling, your lungs expand and air pressure in your alveoli drops. Atmospheric air rushes in to make up the difference. When exhaling, the lungs contract and the alveoli get squeezed. Air pressure rises, and air rushes out to bring down the pressure. This is because of **Boyle's Law**, which says that the pressure and volume of a gas are inversely related.

Isn't It Amazing!

Mammals have tiny alveoli that are densely packed into their lungs, to increase surface area. For example, while a frog's alveolus is 10 times as wide, it has only 20 cm^2 of gas exchange area per cc, while human beings pack in 300 cm^2 of gas exchange area per cc.

▲ *Frogs do most of their breathing through their skin*

The Respiratory Cycle

Your doctor's term for breathing is pulmonary ventilation (PV) or a respiratory cycle. One PV has two steps—breathing in or inspiration and breathing out or expiration. Inhalation is the term for when you deliberately breathe something in, like the air above food that smells delicious. Exhalation is for breathing out. Most of the time, you breathe quietly. This is called **eupnea**. When you are tired or doing something that needs more breath—like singing—your body switches to forced breathing. This is called **hyperpnea**.

Singers need to control their breathing to carry a tune properly

You use two sets of muscles to breathe—the muscles between your ribs, called **intercostal muscles** and the diaphragm, that is the large muscle stretching between the ribs, sternum, and lumbar vertebra.

Trachea
Rib cage
External intercostal muscles
Lungs
Bronchial tubes
Pleura
Diaphragm
Diaphragm
Inspiration
Expiration

Respiration works by using Boyle's Law of gas pressures

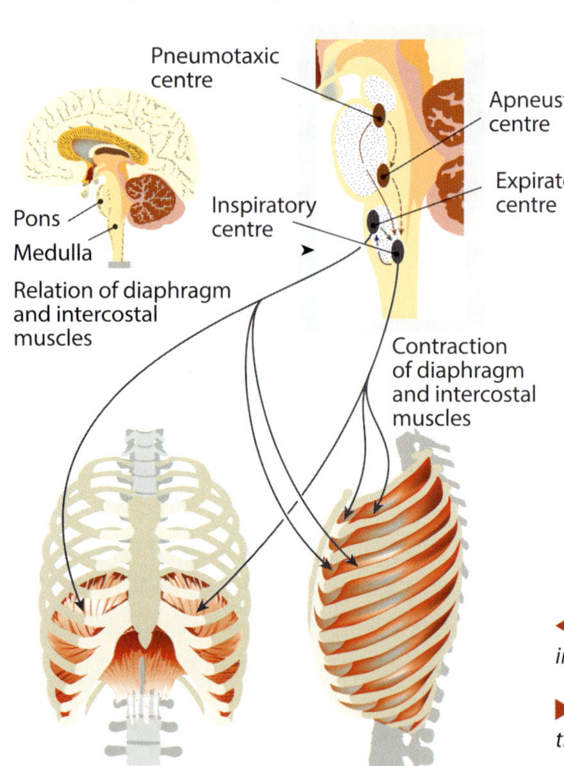

Quiet Breathing

Quiet breathing may be shallow or deep. In the shallow kind, only the intercostals contract and expand. Deep breathing involves the diaphragm too. During inspiration, the diaphragm contracts and pulls the pleurae down. The intercostals contract and pull the ribcage and the attached pleurae, upward and outward. Together, they expand the lung and air rushes in to equalise the alveolar pressure. When the muscles relax, they squeeze the pleurae, which then squeeze the lungs, forcing air out.

In quiet breathing, your skeletal muscles work like involuntary muscles, controlled by the brain stem

Forced breathing helps get rid of the lactic acid built up by the muscles during exercise in the form of CO_2, by oxidation

Forced Breathing

Other muscles join in when you need to breathe in or breathe out more air. This happens when your body needs more oxygen, like when you are tired after exercise or heavy work, or if you experience shortness of breath, or if you have asthma.

When breathing in, the neck muscles pull the ribcage upward, enlarging the chest beyond what the intercostals and diaphragm can do. Put your fingers on the base of your neck and breathe in deeply. Can you feel your shoulders rise?

When breathing out, the muscles of your belly, like the obliques, contract. This squeezes your belly, and the organs in it push the diaphragm up in turn. Pull in your stomach, and you can feel air rushing out of your nose.

Taking Oxygen to the Tissues

Oxygen diffuses from the alveoli into the pulmonary capillaries under atmospheric pressure. But oxygen molecules are not easily soluble in liquids or water, which is what most of your blood is. Only 1.5 per cent of all the oxygen in your blood is actually dissolved. The rest has to be literally carried through the blood to the tissues. That job is done by the molecule haemoglobin, which is present in red blood cells.

Haemoglobin

Haemoglobin is made of three parts—two protein parts called alpha-globins and beta-globins, and an organic molecule called haeme. At the centre of each haeme molecule is an iron atom. The chemistry of haemoglobin makes the iron in the haeme very attractive to oxygen. As blood enters deep inside the lungs, the haemoglobin molecules take up oxygen very quickly to become **oxyhaemoglobin**.

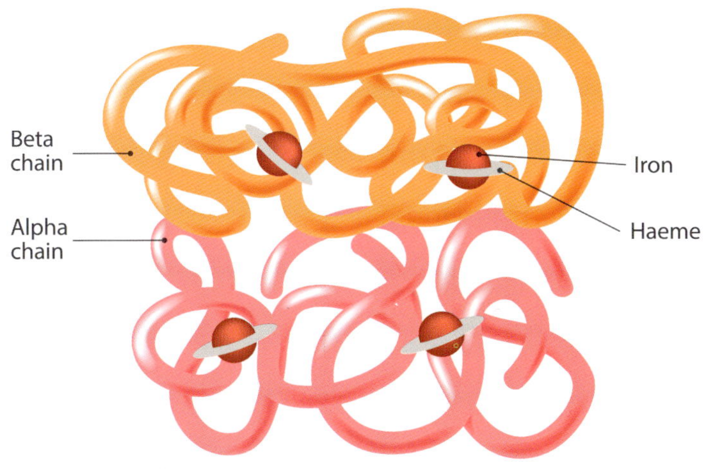

▲ *It is the iron atom that gives haemoglobin, and therefore blood, its bright red colour*

Delivering Oxygen

When oxygenated blood reaches the tissues, it reaches a place low in oxygen and high in acidity. The chemistry of haemoglobin now changes and oxygen is released. It diffuses from the plasma, into the cells, where it is taken up for respiration.

Haemoglobin and oxygen break up faster under higher temperatures. During exercise, muscle tissues release a lot of energy and heat up. Haemoglobin going to the muscles thus gives up oxygen more easily.

Carbon Monoxide

Haemoglobin is attracted even more to carbon monoxide (CO) than to oxygen. CO is a colourless odourless gas that is often found in the exhausts of vehicles and central heating furnaces. It forms carboxyhaemoglobin in the blood, which does not break up in the tissues. When a person inhales CO, they get headaches, feel dizzy, and have pain in the chest. They can die quickly as their tissues, especially those in the brain, die of oxygen starvation. If you suspect a person of having CO poisoning, switch off all flames and electrical devices and take them outdoors.

Incredible Individuals

John Scott Haldane (1860–1936) was a scientist concerned with the safety of coalmine workers, many of whom died of poisoning deep in the mines. He discovered that this was because of carbon monoxide. He published a report on safety in mines which got translated into many languages. He also designed a respirator that would prevent gas poisoning and studied the effect of very high altitude on breathing. (*see pp 18*)

▲ *Carbon monoxide is called the silent killer because it can neither be seen nor smelled*

Cellular Respiration

The word respiration has two meanings. When doctors and scientists talk about the whole body, they mean inspiration and expiration. But when they talk about each cell of your body, respiration has a different meaning. It means how your cells convert their main fuel, that is glucose, into energy. When glucose is chemically broken down by the enzymes in your cells, it releases heat. This heat is used to make another chemical called Adenosine Triphosphate (ATP). Whenever the body needs quick energy, like when you exercise, the ATP is broken up and the heat stored is released again.

▲ *Make sure you are breathing well while exercising*

Glycolysis

The first step of respiration is glycolysis, in which glucose is broken in a series of reactions into a smaller compound called pyruvate. If the cell has time, it will then carry out aerobic respiration. If the cell is in a hurry, as muscles cells are when you are running, it will carry out anaerobic respiration.

Aerobic Respiration

The oxygen you breathe in is used for aerobic respiration, in which glucose is broken down completely into carbon dioxide and water. Pyruvate goes through the Citric Acid Cycle (also called Krebs Cycle), a series of chemical reactions that turn it into NADH and $FADH_2$. These react with oxygen to form ATP and release CO_2.

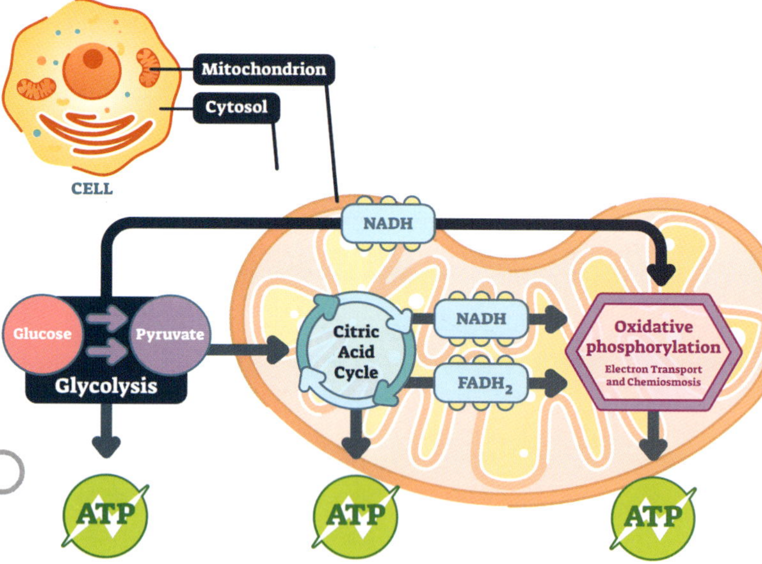

▲ *The diagram shows the process of aerobic respiration*

Anaerobic Respiration

When the cell needs energy fast, it only carries out glycolysis and the pyruvate is turned into lactic acid instead. Less ATP is made, so you run out of energy and feel tired. The lactic acid is sent to the liver, where it is turned back into glucose, while you are resting.

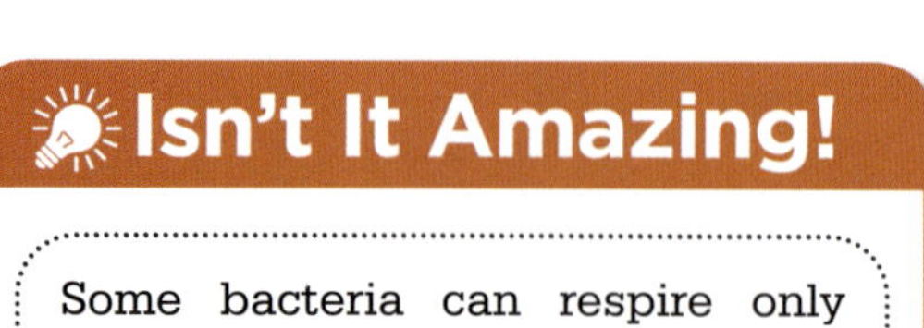

Isn't It Amazing!

Some bacteria can respire only anaerobically. They are useful to us in making cheese and yoghurt from milk. The lactic acid that is produced makes milk congeal. This is called fermentation.

▼ *Cheeses made around the world are made by the process of fermentation*

Your Body's Sound Box

Just below the epiglottis, where the trachea begins, is your larynx. It is also known as the voice box. Made of cartilage and muscles, it has an interesting structure that allows you to make sounds. Human beings have among the most complex larynges, which allow them to make dozens of different sounds. They can string the sounds together to make words and sentences. This includes all the grunts and hisses you make when you are too busy to reply properly to someone. But did you know that your larynx develops from the same tissue that, in fish, becomes gills? Such organs are called evolutionary homologues.

Structure

If you feel your throat, you will realise it is hard, though not as hard as a bone. This is where the voice box is, made of three pieces of cartilage—the epiglottis, thyroid cartilage, and cricoid cartilage. The thyroid cartilage is bigger in men and can be felt easily. It is called Adam's apple. The cricoid cartilage is thicker and ring-shaped. Three small pairs of cartilage are attached to the epiglottis—the arytenoids, corniculates, and cuneiforms—that help move the vocal cords.

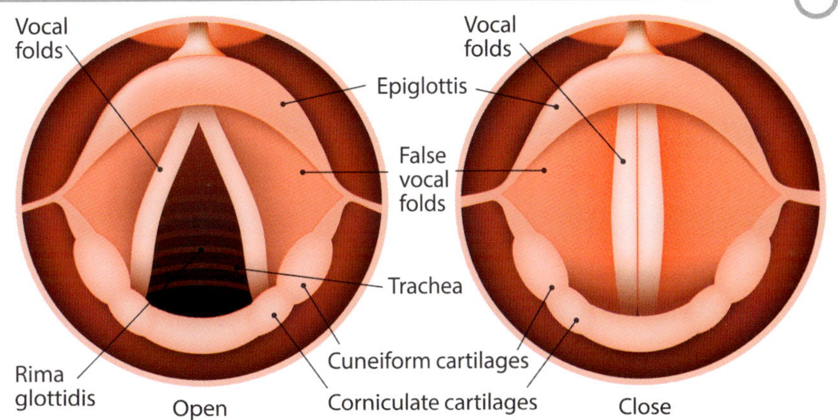

Below the epiglottis is the glottis, made of folds of tissue called the true vocal cords and the false vocal cords. True vocal cords are thin and membranous, held in place by the cartilages and vocal muscles. As air passes, they vibrate like guitar strings to make sound. Men have broader cords that give them a deeper voice, while women have narrower ones, so they sound shriller than men.

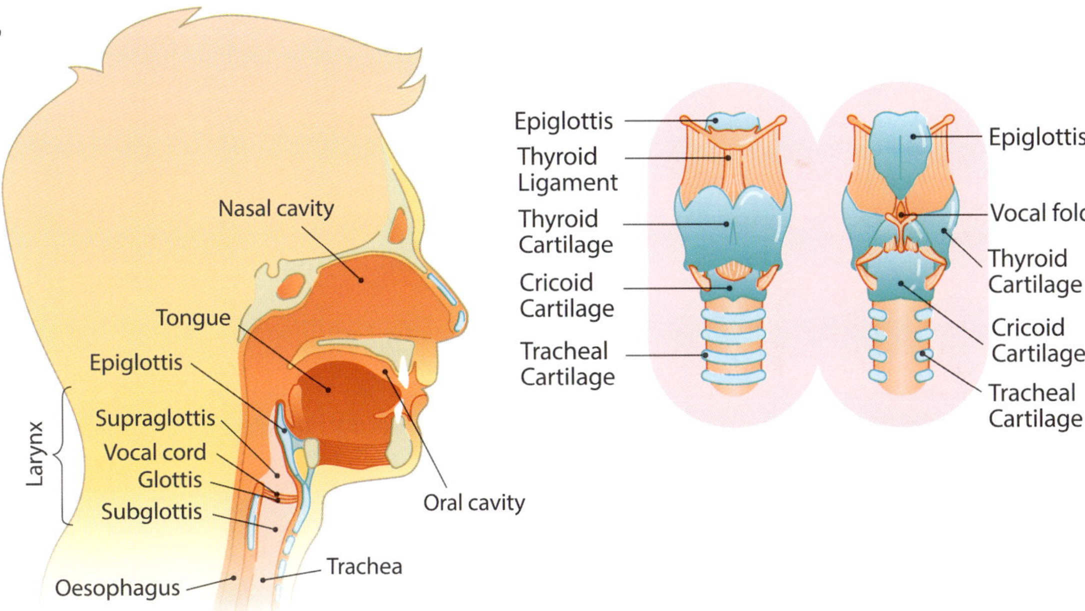

▲ *The vocal folds close when we speak, so that the air going out makes them vibrate*

In Real Life

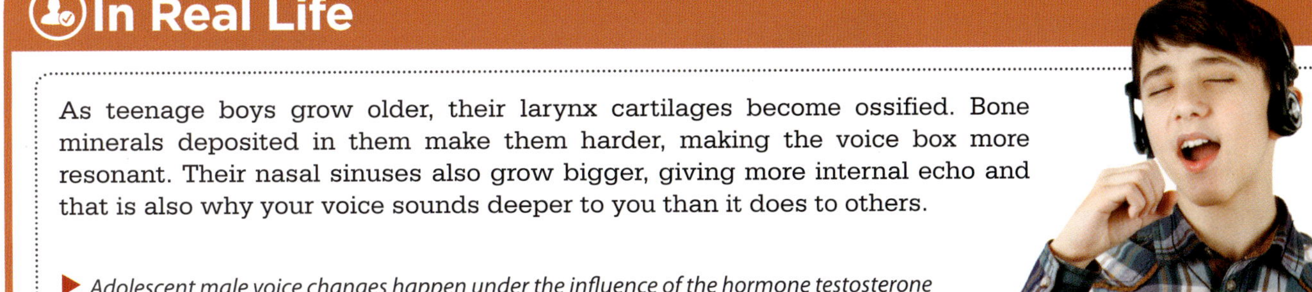

As teenage boys grow older, their larynx cartilages become ossified. Bone minerals deposited in them make them harder, making the voice box more resonant. Their nasal sinuses also grow bigger, giving more internal echo and that is also why your voice sounds deeper to you than it does to others.

► *Adolescent male voice changes happen under the influence of the hormone testosterone*

Protecting the Brain: The Skull

The skull has fascinated people from the beginning of time. It is a symbol of death and danger in many cultures. This could be attributed to the fact that the brain and the head are considered to be representative of a person. Brain damage can seriously impair one's life, and the functioning of any and all other parts of the body.

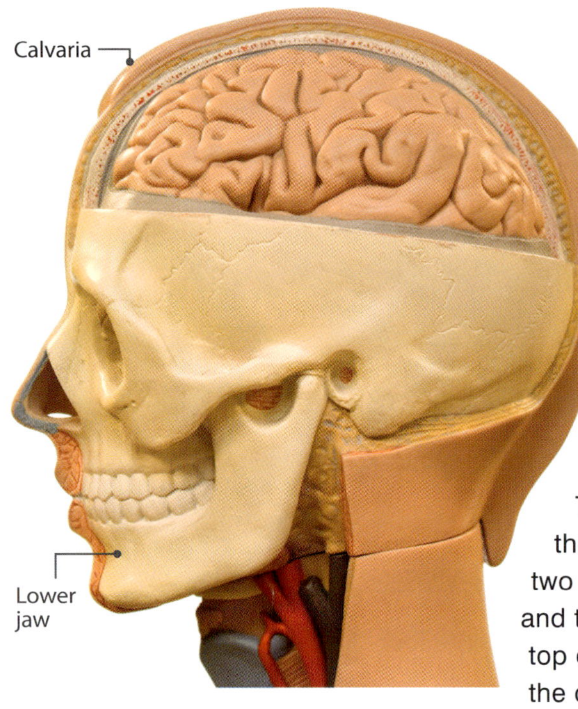

The Brainbox

Human life wouldn't be the same without the skull. It protects the brain and the organs of the face, such as the eyes, nose, inner ear, and mouth. It is harder than most bones, so if you have a bad fall, your skull does not fracture easily.

The skull, also known as the cranium, is made of two parts—the brainbox and the facial bones. The top of the skull is called the calvaria, its sides are called temples, while the bottom is called the base. Large plate-like bones enclose the brain. They are the frontal bone—the forehead, a pair of parietal bones—at the back of your head, a pair of temporal bones—on the sides, and the occipital bone—at the back of your neck. They correspond to the lobes of the cerebrum.

The temporal bone has a hole to let in the ear canal and the mandibular fossa, into which the lower jaw fits. The occipital bone has a large hole in it, through which the spinal cord passes from the brain to the body. The sphenoid makes the floor of the brain box. It is full of small holes to let in cranial nerves and arteries and let veins out. The last is the ethmoid bone, which also makes the roof and septum of the nasal cavity.

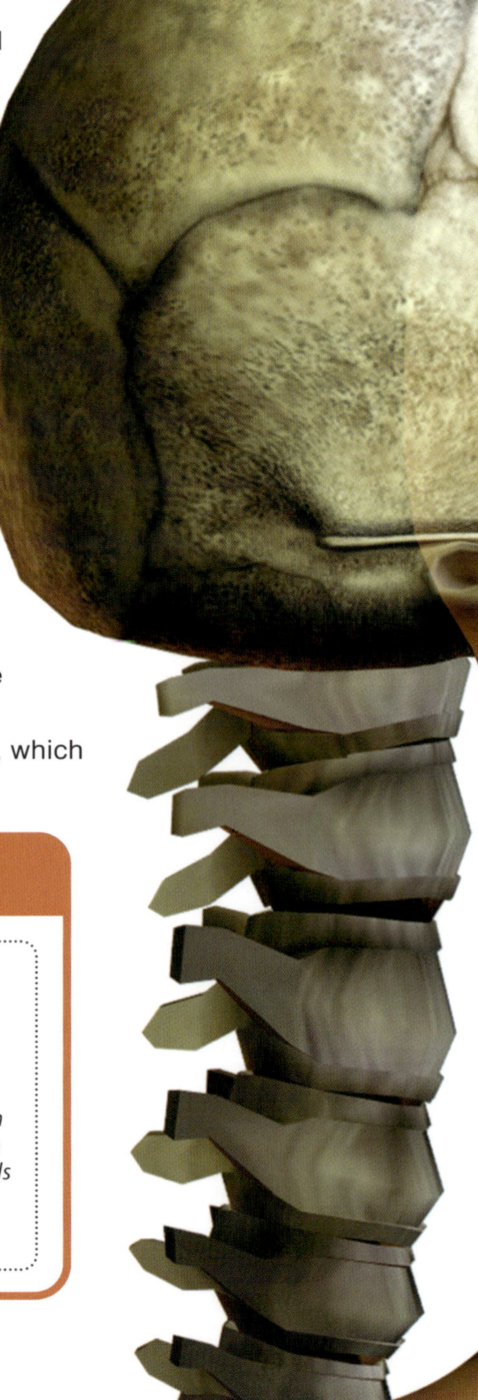

In Real Life

The human brainbox—in proportion to the rest of the body—is among the largest in the animal kingdom. It has to be, for it evolved to have a brain that is also the largest in proportion to the body. The brain takes up over two-thirds of the skull.

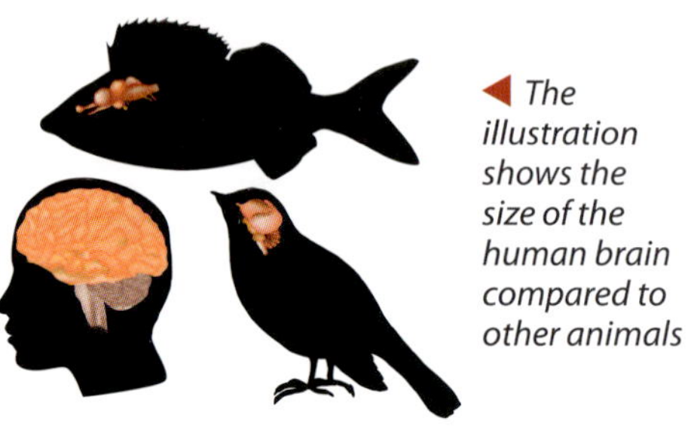

The illustration shows the size of the human brain compared to other animals

Incredible Individuals

Have you seen someone in a plaster cast? They might have needed it because of an injury that broke their bones. This technique of using plaster to grip broken bones in a plaster cast was invented by a Dutch army doctor named Antonius Mathijsen (1805–1878) in 1851. He realised that gypsum plaster, when mixed with water, would set very quickly, keeping bones in place.

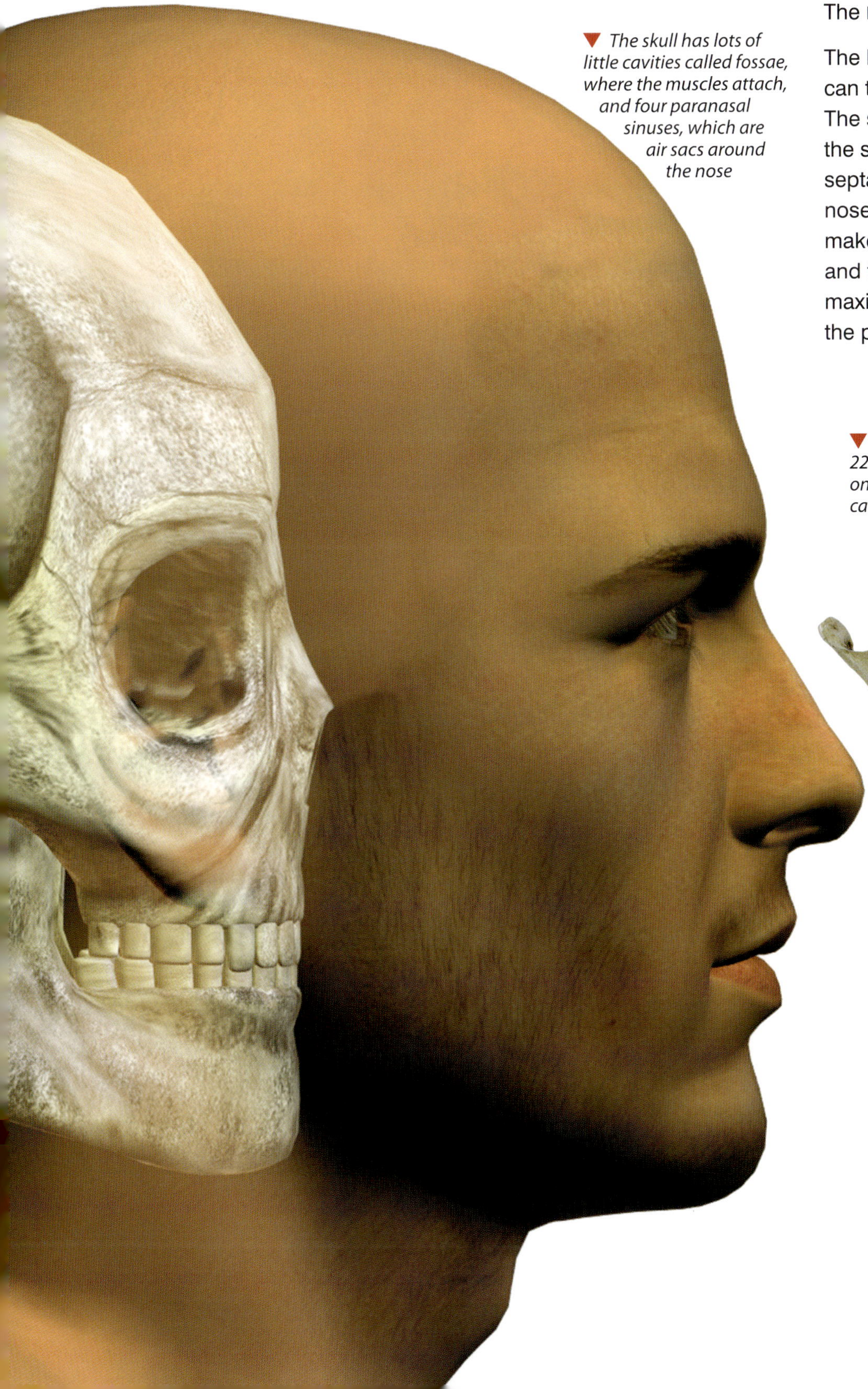

The skull has lots of little cavities called fossae, where the muscles attach, and four paranasal sinuses, which are air sacs around the nose

Eyes and Nose

Your eyes sit in two cavities in the skull called the **orbit**. The frontal bone makes up the upper half of both the orbits. Your cheekbones, that is, zygomatic bones, make up its outer sides. The optic nerve passes through the sphenoid in their back. The tiny lacrimal bones, and parts of the ethmoid, make up the inner sides. The **maxilla** makes up its floor.

The bony part of the nose that you can touch is made of the nasal bones. The single vomer and the ethmoid make the septum, which divides the nostrils. The septal cartilage, which makes your visible nose, attaches to the ethmoid. The palate makes both the floor of the nasal cavity and the roof of the mouth. It is made by the maxilla, and a small pair of bones called the palatines.

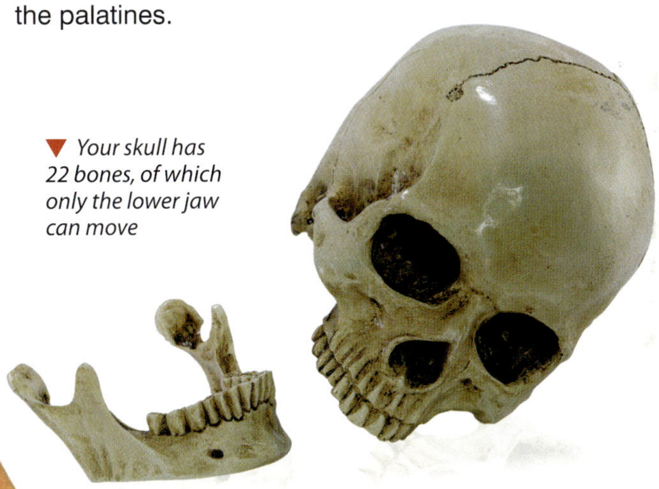

Your skull has 22 bones, of which only the lower jaw can move

The Mouth

Your mouth is made of the jaws and the palate. The jaws contain your teeth. There are 24 milk teeth in children and 32 teeth in adults. Your upper jaw is called the maxilla, which is actually a pair of bones joined together. The lower jaw is called the **mandible**. It is attached to the temporal bones on both sides of the mouth. The hyoid is a special bone, which is not attached to any other bones. It is behind the mandible and the tongue attaches to it.

Stiffening the Back
The Vertebral Column

▼ *Let's take a look at the vertebra*

The vertebral column, also called the spine, divides the animal kingdom into two—those without it—invertebrates and those with it—vertebrates. Fish, amphibians, reptiles, birds, and mammals are vertebrates, with a bony spine. In some fish, like sharks and rays, the spine, like the rest of the skeletal system, is made of cartilage. It develops from the notochord, a stiff rod-like organ in the embryo.

Spine

The spine is made of 33 bones called vertebrae (singular: **vertebra**). It protects the spinal cord, and provides an anchor for the ribs, the shoulders, and the hips. It also helps you flex your body in many ways, so you can run around and play with ease.

The spine has five regions. As you grow, the vertebrae of the sacrum merge into each other to form a single bone which becomes part of the hip. The four tail vertebrae also merge to form the tiny coccyx, which curves inwards. Animals with tails have many more vertebrae in them.

Sometimes the gel-like part of the intervertebral disc gets crushed and comes out of the spine. This is called a slipped disc. It causes a lot of pain.

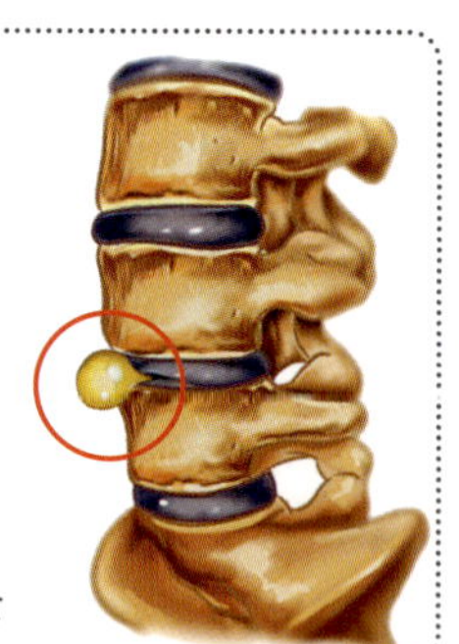

▶ *An illustration of slipped disc*

Region	Type of Vertebrae	Number	Names of Vertebrae	Attaches to
Neck	Cervical	7	C1–C7	The skull
Upper Back	Thoracic	12	T1–T12	The ribs
Lower Back	Lumbar	5	L1–L5	The sacrum
Hip	Sacral	5	S1–S5	The hip bones
Tail (Coccyx)	Coccygeal	4	Tailbone	Bottom of the sacrum

Vertebra

Each vertebra has two parts—an anterior segment or the vertebral body, and a posterior part or the vertebral arch. Each vertebra has 'processes', which is an outgrowth of tissue or cell. Between the bodies of two vertebrae are **intervertebral discs,** made of cartilage. The outer part of each is fibrous, while the inner part is jelly-like. These discs act as cushions when the vertebrae bend backwards or forwards, so that their bodies do not rub against each other. The vertebral arch extends in two arms that meet behind the vertebral foramen. The foramina of all the vertebra form a tube-like canal, through which the spinal cord passes.

The segmentation of the spine gives your body incredible flexibility

Finally, the processes act as places where skeletal muscles attach. Vertebrae in different regions of the spine have different processes. The spinous process sticks out of each vertebra and points downwards. This stops you from twisting yourself too much and dislocating your spinal cord. The gaps between the vertebra are also spaces from which the spinal nerves pass from the spinal cord to various parts of the body.

The S-shape of the spinal cord allows us to walk and run

Regional Vertebrae

Cervical vertebra has extra foramina, through which the arteries to the brain pass. The first of them (C1), is called atlas. It bears the weight of the skull. Its body is hollow. The second (C2), is called axis. Its body has a pivot called the dens, which fits into the hollow of atlas. This is what lets you turn your neck around.

The five regions of the spine

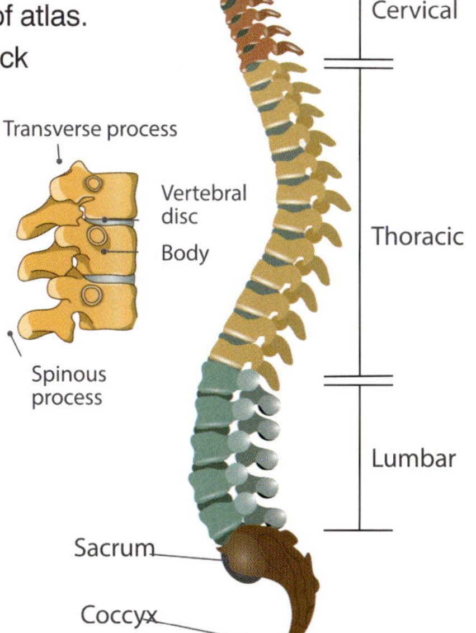

The thoracic vertebra has processes to the side called transverse processes. Both they and the vertebral arch have dimples on both sides called facets. Here is where the ribs attach. Lumbar vertebrae have additional processes called articular processes, where muscles can attach. The fused arches of the sacral vertebrae make up an ear-like surface, to which the hip bones attach.

Isn't It Amazing!

The bony plates of a stegosaurus were not part of its spine but grew from the skin covering the spine.

Skeleton of a stegosaurus

Inside Your Bones

Our skeleton comprises our bones. But what exactly exists inside each bone? Even though they often feel like nothing but hard, white rods, bones, in fact, have a complex internal structure. Not all bones are the same. Some are much harder than the others, like the skull plates. Other bones are spongy and lighter than they feel. Many bones are hollow inside. However, they are filled with a fatty substance called **marrow**. Did you know that this marrow is the place where your blood cells are born?

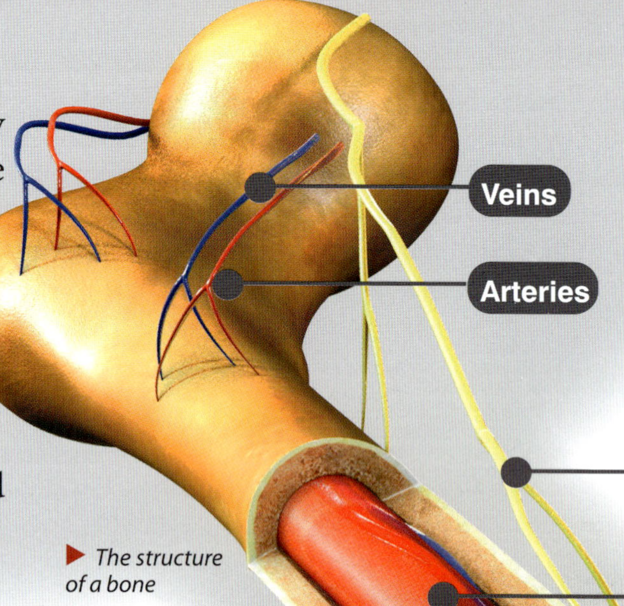

The structure of a bone

The Making of a Bone

We start out as babies with a cartilage skeleton. Each 'bone' has a tissue that covers it, called the periosteum. Special cells, called osteogenic cells, migrate to the bone being formed. Some stay as they are until the day that the bones need to be repaired. Most will become **osteoblasts**, which are cells that make **collagen**. Collagen is a protein that forms threads easily, and is seen in **tendons** and **ligaments**, as well as in hair and nails. In the bones, it forms a three-dimensional 'matrix' of woven fibres.

Osteoblasts fill this matrix with calcium hydroxyapatite, a mineral compound made of calcium and phosphorus ($Ca_{10}(PO_4)_6(OH)_2$). The bone matrix may be filled completely with this compound to make compact bone, or leave air gaps to make spongy bone. As the osteoblasts lay down minerals, they get trapped in it and become osteocytes. Osteocytes keep the bone nourished and lay down more mineral if needed.

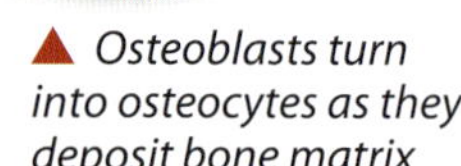

Osteoblasts turn into osteocytes as they deposit bone matrix

As we grow, so do our bones. Special cells called **osteoclasts** help in this process. They 'resorb' the minerals so that the bone can become wider and longer, before osteoblasts lay down fresh matrix.

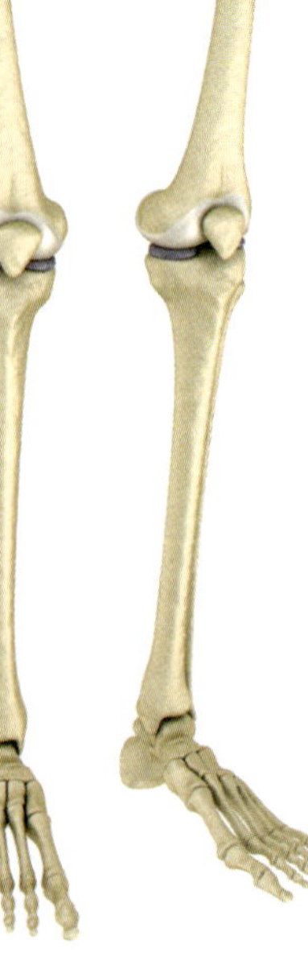

Bones do not stay the same throughout life but are constantly remodelled to grow in size and weight as you grow

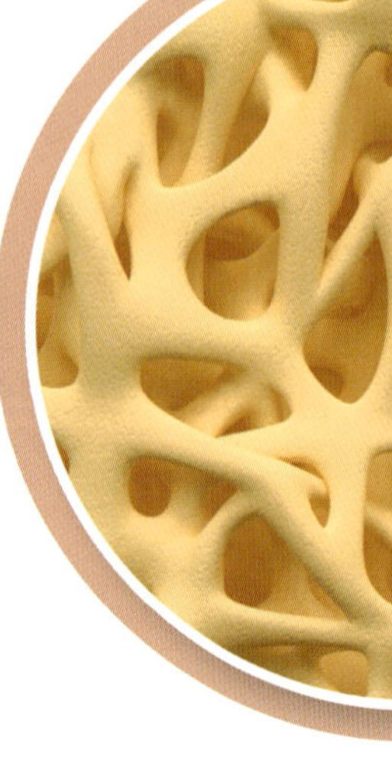

Isn't It Amazing!

Did you know that shark bones are actually made of cartilage? Cartilage is a light, flexible material made of chondroitin sulphate, hyaluronic acid, and collagen fibres. It is also found where bones meet to form a joint.

Sharks are called cartilaginous fish

Incredible Individuals

Scientists believe that the ancestors of our species (*Homo habilis*) first started out as scavengers. They ate meat left behind by other predators. They started making tools out of stone to get to the marrow of left-behind bones, and that was one of the first tools that humankind made.

Bone Marrow

Many bones, like the vertebrae, ribs, sternum, hip bones, and bones of the arm and leg are hollow inside and filled with either yellow or red marrow. Yellow marrow is a jelly-like tissue made of fat-storing cells. Red marrow is more complex and is made of stem cells that make the cells of the blood—Red Blood Cells (RBCs), White Blood Cells (WBCs), and platelets.

Until you are seven years old, almost all your marrow is red. After that, most marrow becomes yellow and stops making blood. But if a bad injury or fever with lots of blood loss occurs, yellow marrow can become red again.

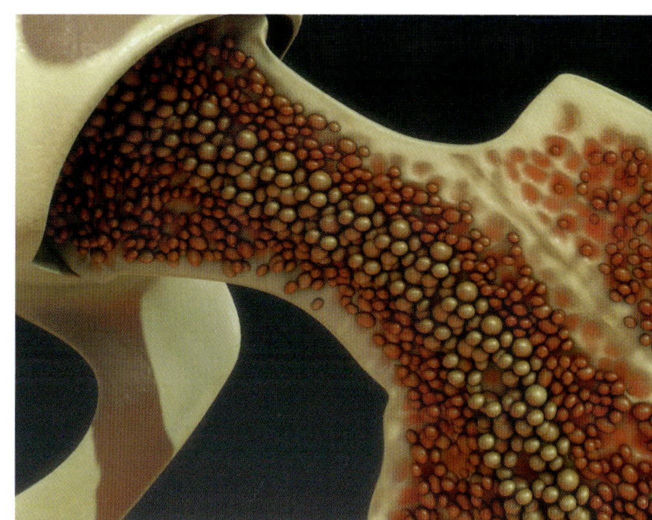

▲ *A 3D illustration of bone marrow*

◀ *Normal bone matrix*

Bone Membranes

The inside of a bone is called its medullary cavity. It is lined by a membrane called the endosteum. It contains the bone cells that help the bone grow and also repair it. The outside of a bone is also covered by a membrane called the periosteum. Blood vessels, nerves, and lymphatic vessels come to the periosteum, from where nutrients diffuse into the bone. Tendons and ligaments attach at the periosteum. At the joints, the periosteum is replaced by articular cartilage.

How Joints Work

A skeleton may appear creepy, but observing it will help you understand how our bodies move. The joints you see here are all synovial diarthroses, meaning they allow plenty of movement and are lubricated by synovial fluid.

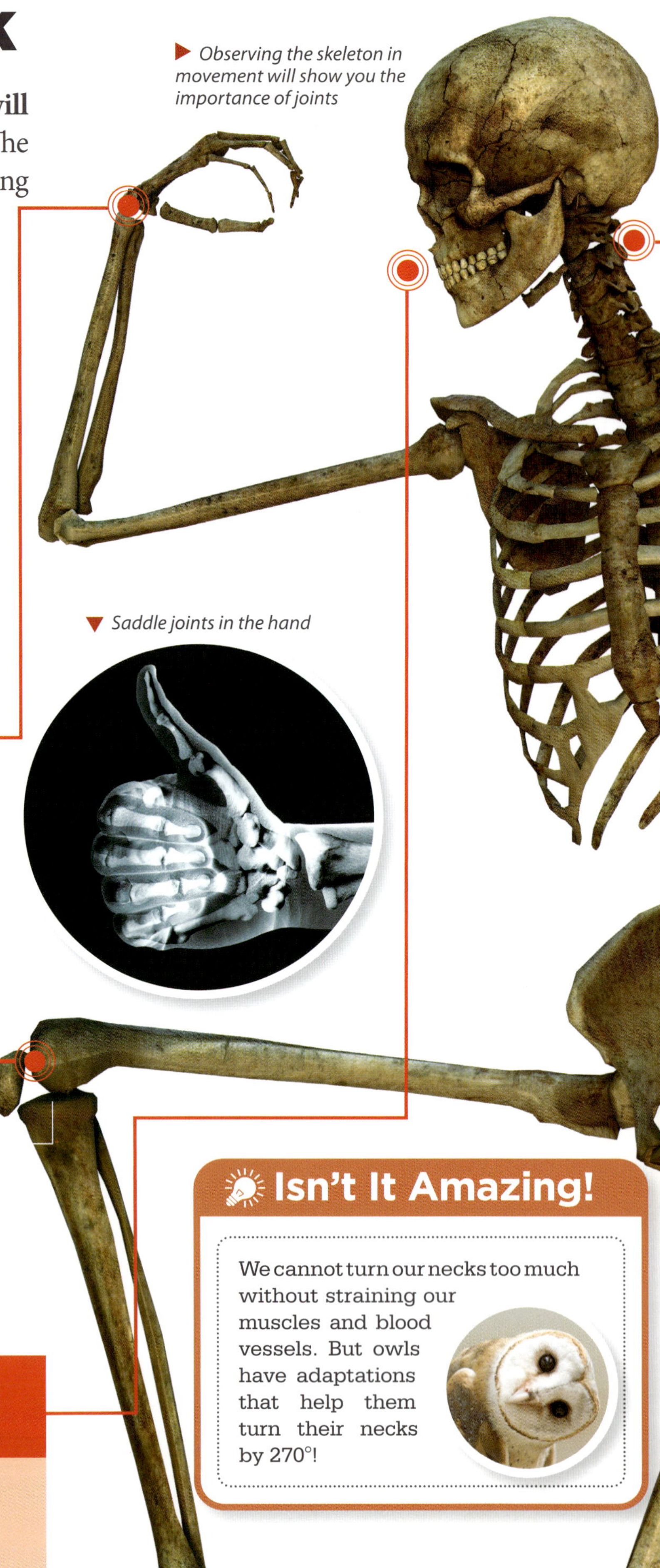

▶ *Observing the skeleton in movement will show you the importance of joints*

▼ *Saddle joints in the hand*

More about Joints

Synovial fluid is made of plasma from the blood that is rich in nutrients. It also contains hyaluronic acid, which acts a bit like engine oil. It is elastic and viscous, preventing friction in the joint. **Articular cartilage** covers the ends of the bones, acting like a shock absorber. The rest of the joint is made of the periosteum of both the bones, making a sac that holds the synovial fluid in. So, it is called the synovial sac.

Saddle: Wrist & thumb

The metacarpal of the thumb meets the trapezium in this joint. Both the faces of the bones are saddle-like. Hold your thumb and forefinger apart to stretch the web between them. Bring both hands together at right angles and you can see how this joint works.

Hinge: Humerus & Ulna

The upper end of the ulna is the olecranon process, which locks into the olecranon fossa of the humerus when the arm is stretched fully. This makes sure that the forearm does not bend backward. The patella does the same job in the knee.

Bicondylar: Mandible & Temporal Bones

This joint allows both hinge-like and sideways movement to the extent that the skin can be stretched. You see this best in the joint between the mandible (lower jaw) and the temporal bone, which lets you chew food and talk.

Isn't It Amazing!

We cannot turn our necks too much without straining our muscles and blood vessels. But owls have adaptations that help them turn their necks by 270°!

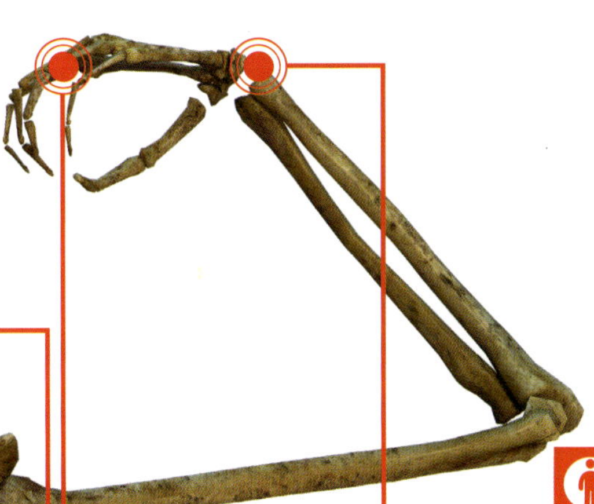

Incredible Individuals

Tommy John's baseball career was finished in 1974 when his elbow ligament was torn. A new surgical procedure got him back in play. He went on to win 164 matches. It is now called the Tommy John Surgery in the USA.

Gliding: Wrist bones

The eight carpals of the wrist have flat faces facing each other. This allows them to glide over each other a bit. Nevertheless, the wrist helps your hand move in all three planes—sideways, up-down and rotational. The tarsals also work this way.

Condyloid & Ellipsoid: Carpal and Forefinger

This joint allows a little movement up and down and a little sideways. You see this in the joint between your palm and your forefinger. It allows all movement except axial rotation. An example of the condyloid joint is the ovoid head of one bone moving into the elliptical cavity of another. This joint is seen in the wrist and the base of the index finger.

Pivot: Axis & Atlas

The dens of the axis—second cervical vertebra—is the pivot onto which the hollow body of the atlas—first vertebra—fits, making a perfect pivot that lets you turn your neck about 100° on each side.

Ball & Socket: Femur and Hip

The head of the femur is shaped like a ball, which fits into the acetabulum of the hip. Several ligaments and muscles help the joint move in three ways—forward and backward (walking), raising and lowering the leg sideways, and rotating the leg horizontally. The humerus and shoulder joint works the same way.

Ligaments

Ligaments connect two bones to each other at the joints. They are made of tough sheets of collagen, so that they don't tear when the muscles stretch out those bones. Proximal ligaments are close to the synovial sac, while remote ligaments attach to their bones further away.

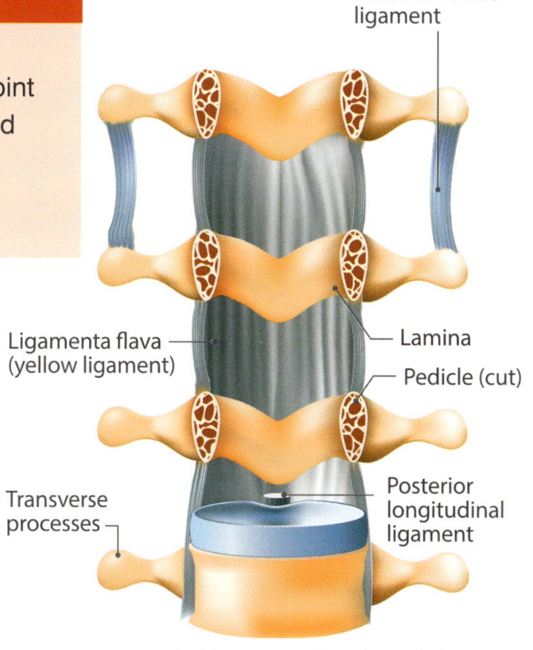

▲ *Ligaments allow for twisting movements of the spine.*

How Muscles Work

Bodybuilders and weightlifters love showing off the bulge in their biceps as they contract. They have exercised a lot for those big muscles, but what are they made of? Why do they appear so strong? How come the muscle bulge is not there when their hands are stretched out or relaxed?

Building muscles also builds and strengthens the bones and tendons

Why Muscle Bulge

Muscles cells contain proteins that do all the hard work. The two main ones are actin and myosin, which form thread-like filaments. Myosin has 'heads', which can stick to actin. When the muscle is at rest, the actin and myosin filaments are apart, and the muscle is flat. When the muscle contracts, millions of actin-myosin pairs are pulled together and the muscle bulges out from the body.

Neuromuscular Junctions

Your muscles are connected to motor neurones through tiny synapses called Neuromuscular Junctions (NMJ). When your muscles have to contract, they get a signal from the brain or spinal cord in the form of a tiny electric current. This current reaches the NMJ, and a chemical called a neurotransmitter is released. It creates a new current across the muscle fibres.

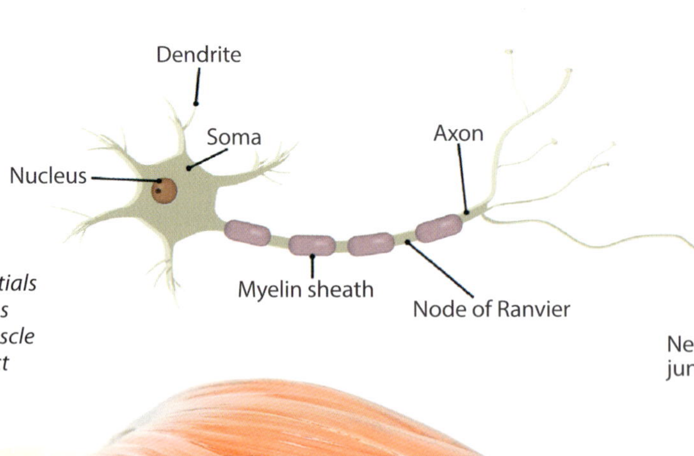

Action potentials from the nervous system tell a muscle when to contract

Action Potential

This is a tiny electric current that passes along the length of a muscle fibre. When there is no message from the brain, there is sodium (Na^+) outside the neurone and potassium (K^+) and chloride (Cl^-) inside the muscle cells. This is the resting potential. When a signal arrives, Na^+ rushes in and K^+ rushes out, creating the action potential. As the muscle relaxes, the two slowly switch places again.

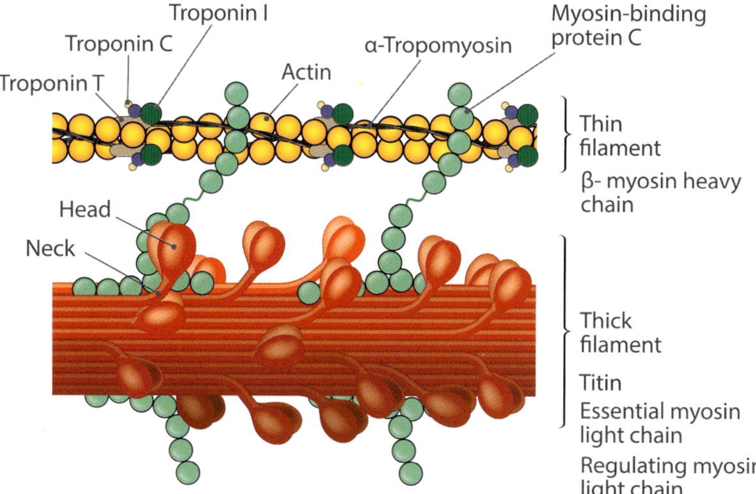

▶ *Lots of different proteins work together to make a muscle contract and relax*

A Muscle in Action

When the brain sends a signal to the muscle to work, it contracts. Like two magnets that stick to each other, the myosin heads and actin filaments stick to each other. The filaments are pulled closer. The sarcomeres become shorter but thicker. As this happens all over the muscle, it bulges out of the body.

Isn't It Amazing!

Muscles work faster at high temperatures and slower at low temperatures. Warm-blooded animals such as birds and mammals maintain a constant body temperature, whatever the weather is, so that their muscles can react quickly to signals from the brain. This helps them run or fly away quickly if faced with danger. Cold-blooded animals like insects and reptiles depend upon the weather being warm to be active.

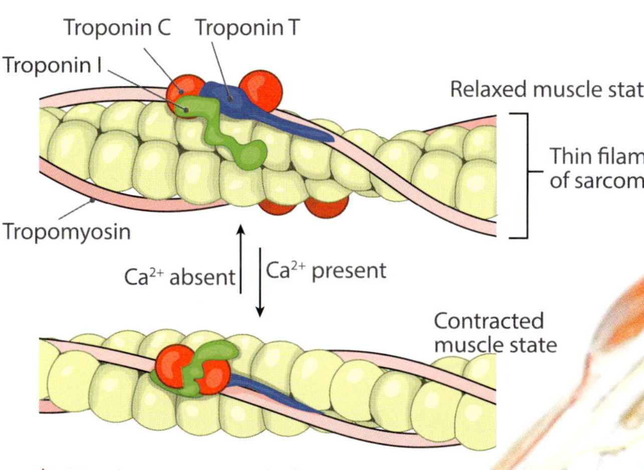

▲ *Muscles contract and relax due to the action of calcium ions on the troponin complex*

Relaxing the Muscle

To relax the muscle again, the body needs to carry out respiration. Enzymes in the muscle cells convert glucose into carbon dioxide and water. This gives out a lot of energy. Some of it is trapped in a molecule called ATP, while the rest goes out as body heat. ATP attaches to the heads of the myosin filaments, causing them to detach from the actin filaments. The filaments come apart, the sarcomeres flatten, and the muscle thus relaxes.

▲ *Yoga can relieve muscular tension*

The Stomach

Think of digestion and you immediately think of the stomach. It controls the pace of digestion and also makes a lot of hormones that communicate with the different parts of the body to either prepare to eat something, or to digest the eaten food and turn it into energy.

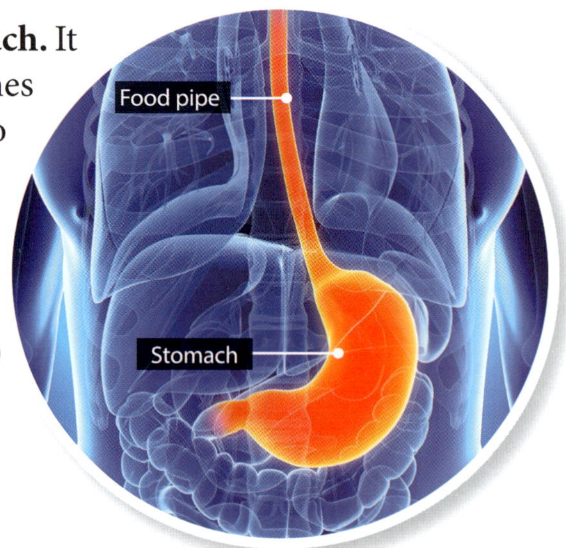

▲ *The food pipe and the stomach*

Getting Food to the Stomach

Between the throat and the stomach is a long pipe that passes between the lungs—the oesophagus, also called the food pipe. But food does not drop down it like water flows through a pipe. The oesophagus has to make the food go down, by squeezing and expanding, peristalsis. Between the oesophagus and stomach is a muscle called the sphincter, which prevents the stomach's acid from getting out.

Inside the Stomach

The stomach is a little like a balloon. It swells up as you eat food. But unlike a balloon, it has very tough walls so that it does not burst. The walls are made of three layers. The outer wall is made of muscles, the middle layer has arteries, veins, and nerves, while the inner wall is made of thousands of tiny folds called gastric pits.

The gastric pits make acid. This acid is released into the stomach when you start eating, and it breaks up the food into smaller chemical bits. The stomach also makes an enzyme called pepsin, which breaks up the proteins that you eat as part of your food into smaller bits called peptides.

The stomach muscles contract and expand when the stomach is full of food. This makes it churn, mixing up the food with the acid and pepsin and speeding up digestion.

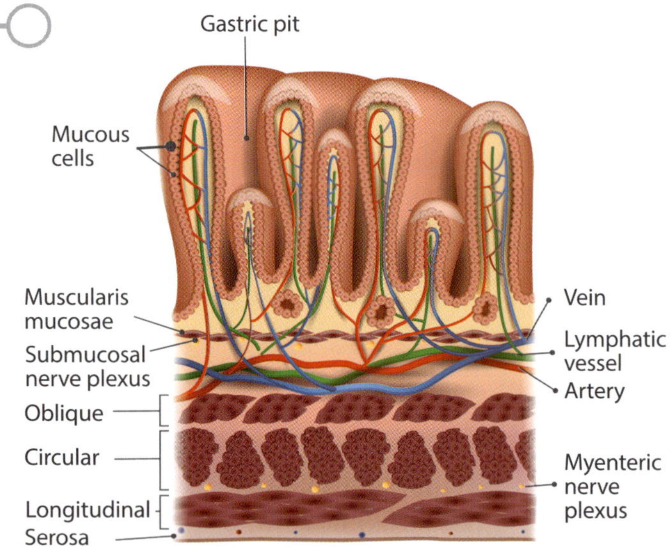

▲ *The stomach does more than just digest food. It also makes many hormones that control digestion and hunger*

Isn't It Amazing!

If you have ever visited a farm, you might have been told that a cow has four stomachs. Actually, the stomach of a cow is so large that it is divided into four parts. This is because the grass and leaves that cows eat are hard to break down and therefore, need a lot of time and space in order for them to be digested.

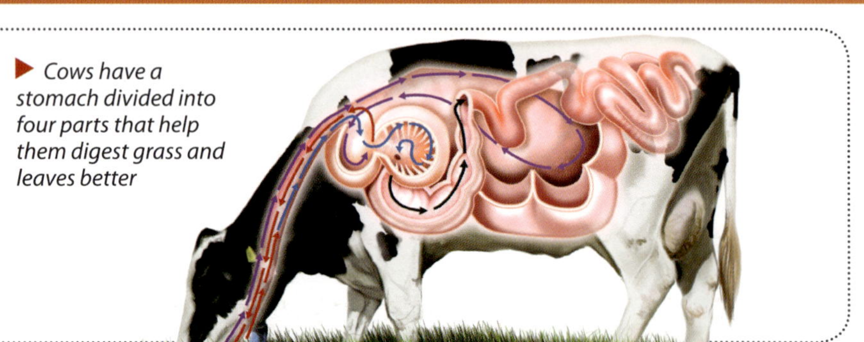

▶ *Cows have a stomach divided into four parts that help them digest grass and leaves better*

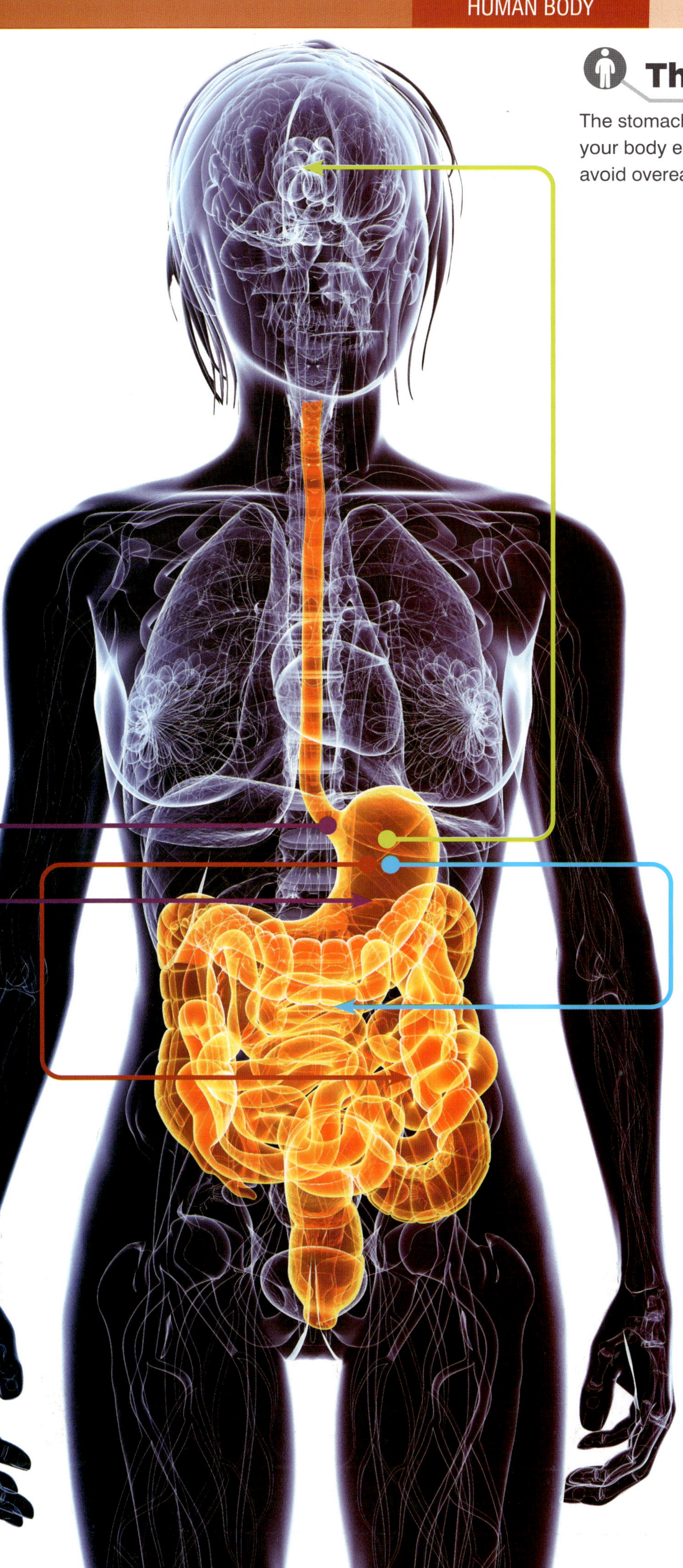

The Hormone Factory

The stomach makes a lot of hormones that help your body eat at the right time, eat enough, and avoid overeating.

- When food enters the stomach, the stomach makes gastrin which tells the gastric pits to release pepsin. It also makes histamine which tells the gastric pits to release acid and serotonin for the stomach muscle to start churning.

- Once you have eaten enough, somatostatin tells the stomach to stop and push the food into the small intestine. It also tells the pancreas and small intestine to stop performing their functions.

- When it is empty, the stomach makes ghrelin, the hormone that tells the brain to feel hungry. The stomach stops making it when food starts entering it.

- Gastrin also tells the small intestine to start its work when new food is coming and the large intestine to throw out the old, digested meal.

◀ *The stomach does more than just digest food. It guides the body to eat healthy*

In Real Life

Somatostatin is also known as the growth hormone-inhibiting hormone. It regulates the endocrine system and is secreted by the D cells of the islets to obstruct the release of glucagon and insulin. It also prevents the release of the growth hormone when it is generated in the hypothalamus. So, there have been many studies conducted on the use of the somatostatin hormone on diseases like breast cancer and malignant lymphoma.

The Intestine

Though it is called the 'small' intestine, it is more than three metres long. It is taller than the tallest human that ever existed! To fit all this length, the small intestine coils and loops around itself, like a long rope would tangle around itself. The small intestine does the hard work of digesting all of your food.

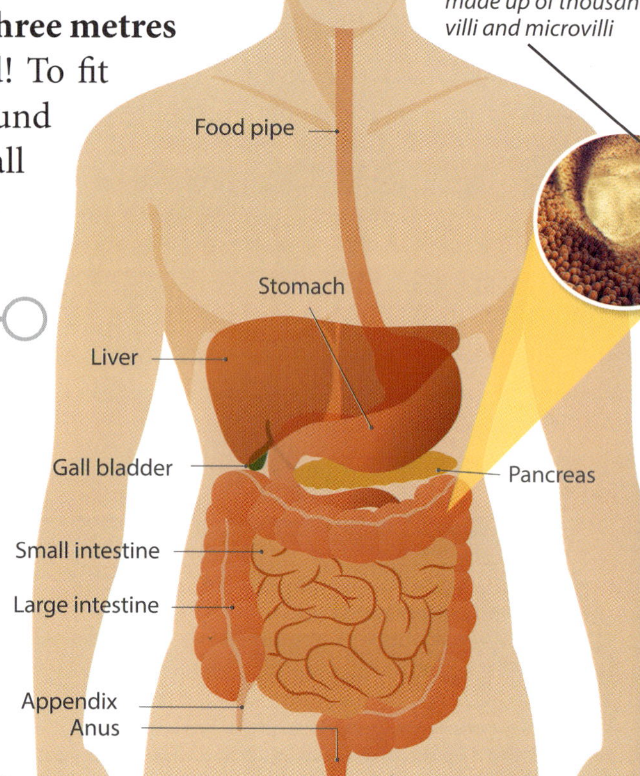

▼ The intestinal wal... made up of thousan... villi and microvilli

The Starter Course: Duodenum

The duodenum is the part that receives the partly digested food from the stomach. Two things happen here:

- Bile from the bile duct acts like soap and makes sure that the fat in your food mixes thoroughly with the rest.
- The pancreas pours a big mix of enzymes that digest the rest of the food. It keeps working till the food reaches the end of the small intestine.

The Main Course: Jejunum and Ileum

The jejunum starts digestion, and the ileum finishes it. The inner walls of both are made of thousands of tiny fingers called villi. Each of these villi are made of even tinier fingers called microvilli. They make the inside border of the intestine very large, so that enzymes can attack every part of the food. The walls of the intestine undergo peristalsis—contract and expand all the time— to keep pushing the food ahead. They also make some more enzymes.

▶ *A microvillus with arteries and veins ready to absorb digested amino acids and sugars*

More Than Just a Digester

The intestine is busy with digestion all day long, but it does other things too. It has tied up with the immune system to create a protective tissue around it called MALT, which keeps bacteria from the intestine escaping into the blood. It also makes a lot of hormones that control digestion, just as the stomach does. One of these hormones is incretin, which tells the pancreas to make another hormone called insulin. Insulin tells the rest of the body to take up the sugar that the intestine just put into the blood.

Another hormone it makes is **serotonin**, which goes to the brain and spinal cord. It tells the nervous system to start the 'rest and digest' system. Serotonin makes the body feel sleepy and convinces the brain that the stomach is full.

Incredible Individuals

Travellers to Latin America often come down with a kind of diarrhoea called Montezuma's Revenge. This happens because of unhygienic hotel rooms or eating street food, where harmful bacteria irritate the intestine. It is named after Montezuma, the last king of the Aztecs who was killed by the Spanish conquerors.

▶ *Montezuma fought wars all over Central America and doubled the size of the Aztec empire*

The Liver

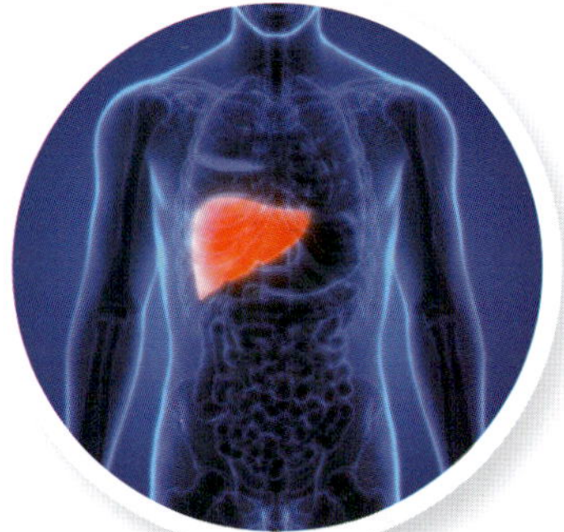

▲ *Location of the liver in the body*

Did you know that the liver is the body's largest gland, weighing nearly 1.4 kg? It has to be, for it is the body's doctor. It is made of cells called hepatocytes.

Storing Glucose

Once you have eaten, your blood suddenly has a lot of sugar in it. Most of this sugar goes to your cells where it is turned into energy. But there is still quite a lot left over. All this goes to the liver, where it turns them into long, stringy molecules called glycogen. If you miss a meal, your brain's hunger centre tells the liver to turn some of the glycogen back into sugar.

Gall Bladder

The liver makes bile all day long, but it cannot store it. Instead, the bile is stored in the gall bladder, where it remains till the brain tells it that food has entered the stomach. Then it pours all the bile into the duodenum, so that the digestion can begin.

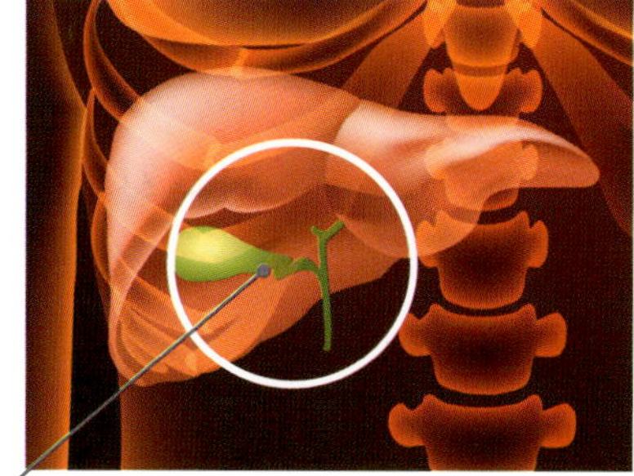

▶ *The gall bladder stores bile until needed. Bile helps in digesting fats*

The Pancreas

The pancreas is not one organ but two. Part of it works for the digestive system by making enzymes, while the rest of it makes hormones that help the body stay healthy and active.

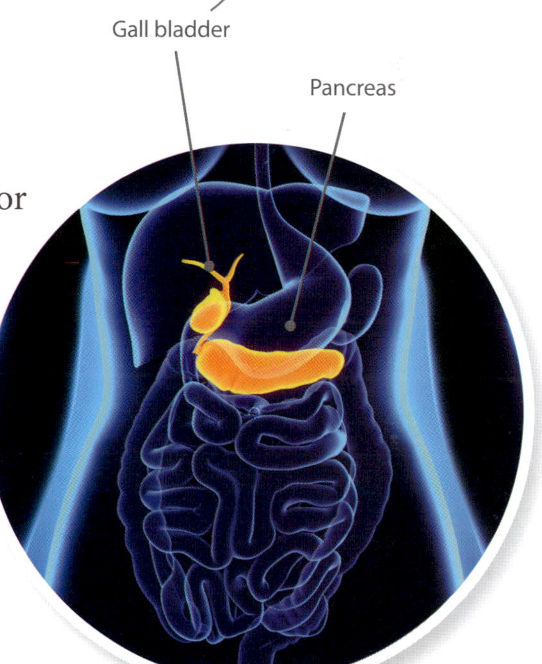

▲ *Pancreas and gall bladder*

Enzyme-making

The cells of the pancreas that make enzymes are called acinar cells. They make many different kinds of enzymes that can digest proteins, DNA, fats, and carbohydrates. The enzymes are collected in tiny pipes that finally flow into the main pancreatic duct.

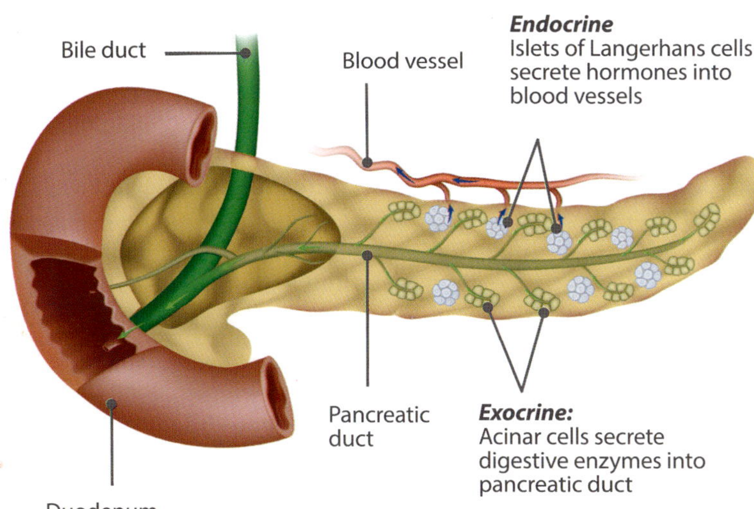

▲ *The pancreas makes both enzymes and hormones. Enzymes are poured into the duodenum, while hormones go into the blood*

Hormone-making

Hormones are made by a different kind of cell, the Islets of Langerhans. The pancreas makes two important hormones:

- Glucagon, which tells the liver to turn glycogen to glucose. This hormone is released when your blood has run out of sugar.
- Insulin, which tells your body's cells to take up sugar from the blood. It is released after you have eaten food.

Enzymes: How We Digest Food

Our body makes several thousand different kinds of proteins that do many things. One kind are enzymes—proteins that help carry out biochemical reactions in the body. Each enzyme catches its own substrate and turns it into the product of digestion—amino acids, simple fats, and sugars.

Enzymes in the Mouth

Amylase is the first enzyme your food meets—in your drool or saliva. It turns starch into sugar. Another enzyme is the lingual lipase, which starts digesting the fat in your food. Your saliva also has lysozyme, an enzyme that breaks up the walls of bacterial cells.

Stomach Enzymes

Pepsin is the main one in the stomach. It breaks proteins into smaller bits called peptides. The stomach also makes gastric lipase, which digests fats. Young children have another called rennin, which helps to digest the protein casein, found in milk and cheese.

Bile

To help you digest the fats you eat, the hepatocytes of the liver make bile. Bile is a yellowish-green soap-like liquid, made of bile salts and bile pigments. Bile salts have two parts—a fatty part that ties up fats in your food and a salt part that mixes them with the intestinal juice. Without bile, the fat would stick to the walls of the intestine and slow down digestion.

Pancreatic Juice

The pancreas makes a lot of enzymes, which it releases together into the duodenum as pancreatic juice.

Enzyme	Acts on	End Product
Carboxy-peptidase	Peptides	Amino acids and peptides
Chymotrypsin	Proteins	Peptides
Elastase	Proteins	Peptides
Pancreatic amylase	Starch	Sugars
Pancreatic lipase	Complex fats	Simple fats
Trypsin	Proteins	Peptides

Enzymes from the Small Intestine

The small intestine's microvilli make the last few enzymes, which finish up digestion.

Enzyme	Acts on	End Product
α-Dextranase	Starch	Glucose
Lactase	Lactose	Glucose and galactose
Maltase	Maltose	Glucose
Sucrase	Sucrose (the most common sugar)	Glucose and fructose
Peptidases	Peptides	Amino acids
Enteropeptidase	Trypsinogen	Trypsin*
Pancreatic lipase	Complex fats	Simple fats
Trypsin	Proteins	Peptides

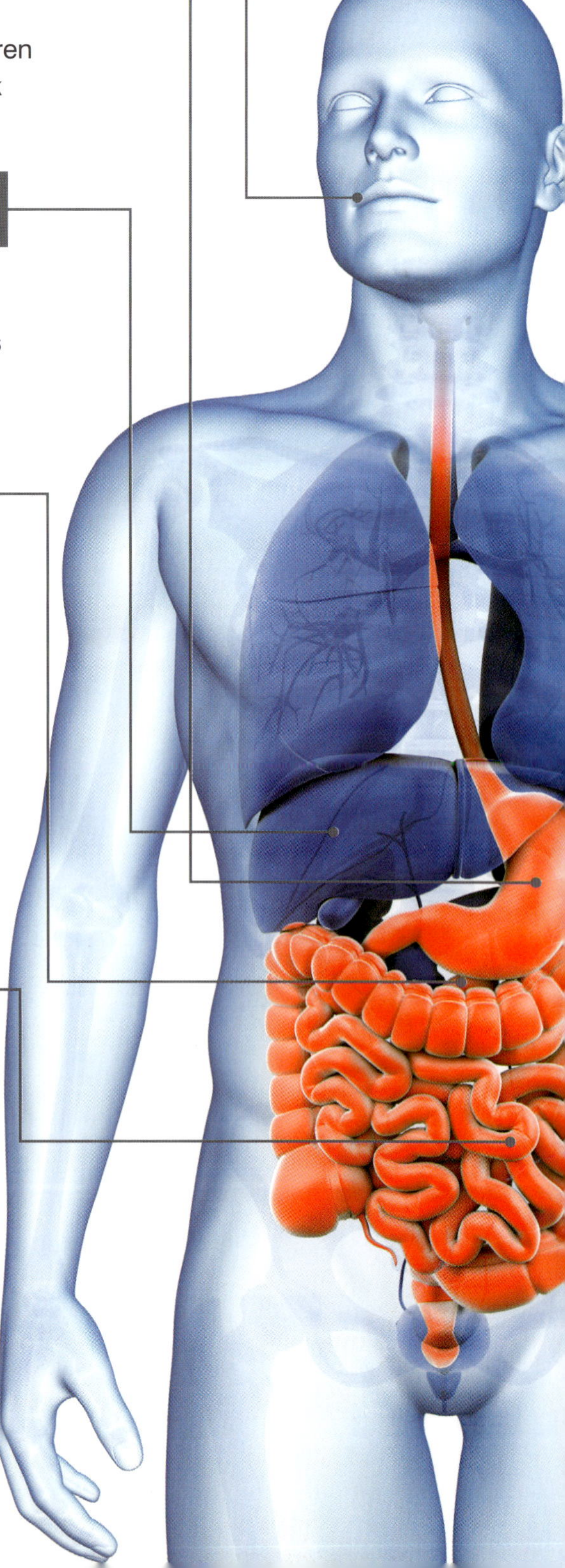

Large Intestine and Appendix

Any picture of the digestive system will show it as a big tube curving around the abdomen. However, most of the food has been digested and absorbed in the small intestine. The stomach and the small intestine also make the hormones that tell the brain to get you to rest while you digest. So, what is left for the large intestine to do? Surprisingly, a lot!

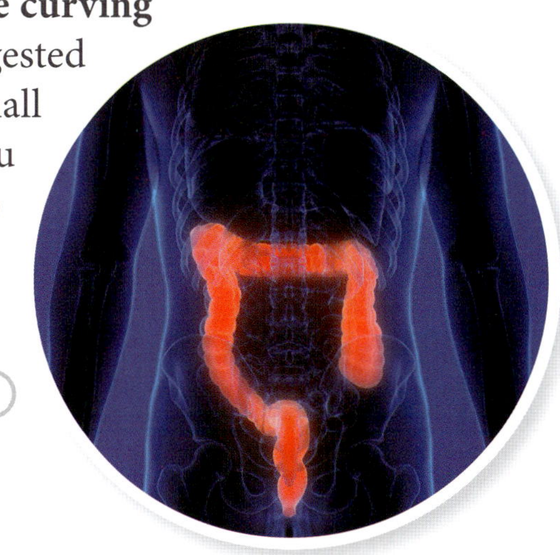

▲ *The large intestine is also known as the colon*

Digesting Water

Did you know that every glass of water you drink, must travel the complete length of your digestive system, before it is absorbed? Why wait so long? That is because your digestive system needs all the water to keep the insides of the stomach and intestines moist till the enzymes have done their work.

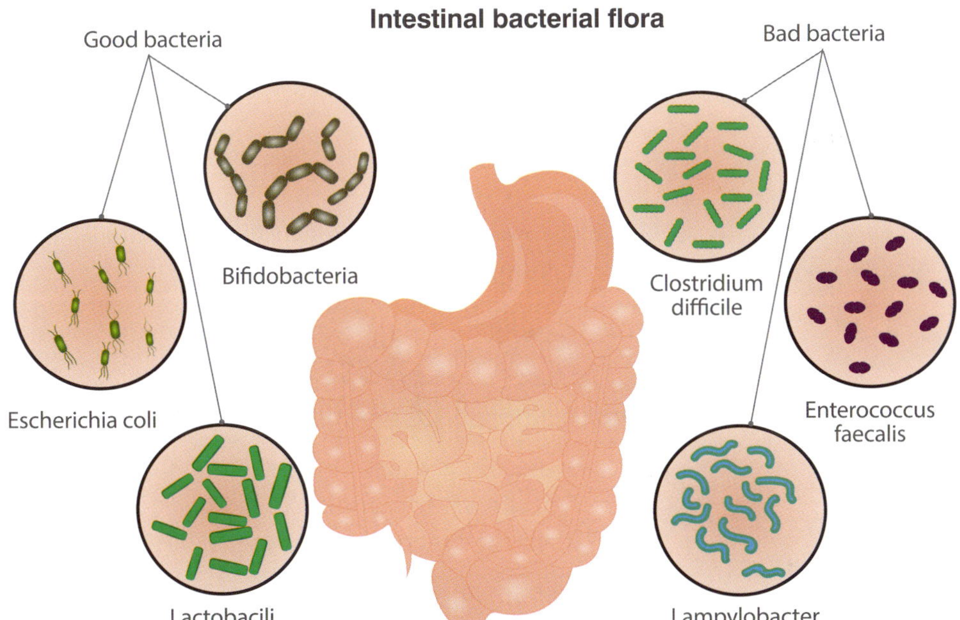

▲ *Good bacteria make vitamins and keep bad bacteria away*

Bacteria

Most of the digestive system is too dangerous for bacteria to live in. But did you know that trillions of them live in the large intestine, belonging to over 700 species? If you remove them, you would lose valuable friends who make some amino acids as well as Vitamins B7, B9, and K; who help you absorb minerals and break down some carbs. They also keep bad bacteria away.

The Mysterious Appendix

This is a tiny part of the large intestine, near where the small intestine meets it. In 2007, some scientists found out that it may act as a training centre for the immune system, and as a refuge for the bacteria that live in the intestine.

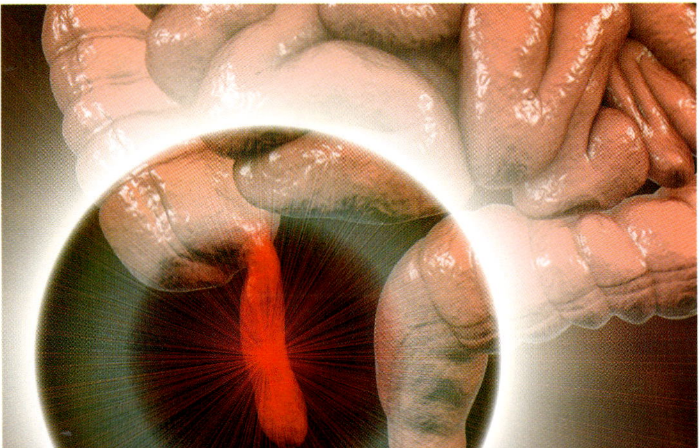

◀ *The painful swelling of the appendix is called appendicitis*

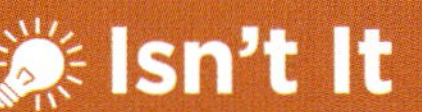

Isn't It Amazing!

A horse's appendix is nearly 4 feet long! In fact, this is true of all plant-eating animals, like cows, goats, and rabbits. The appendix is full of bacteria that help digest the complex carbs found in leaves and grasses.

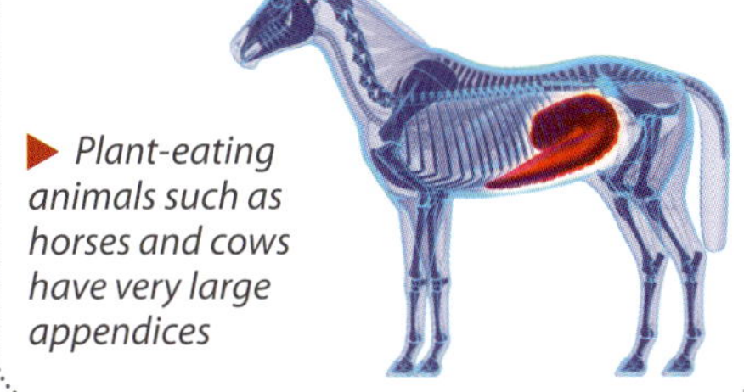

▶ *Plant-eating animals such as horses and cows have very large appendices*

Excretion: Eliminating Waste

Once the food has been consumed and all the protein, sugar, fat, vitamins, minerals, and water from it has been absorbed for the body's use, the remainder is completely unwanted, and must be eliminated. This is carried out through the process of excretion.

Regular exercise coupled with a timely eating and rest regimen helps in excreting wastes at regular intervals.

Keeping Toxins Out

This is actually the body's way of keeping itself safe from infectious bacteria, viruses, and fungi, or from some poisonous things in our food. The small intestine has cells that sniff out toxins. If they find any, the small and large intestinal walls contract very quickly, pushing the food to the anus. It is thrown out, undigested, and watery. Diarrhoea is a common symptom experienced by those suffering from cholera and dysentery.

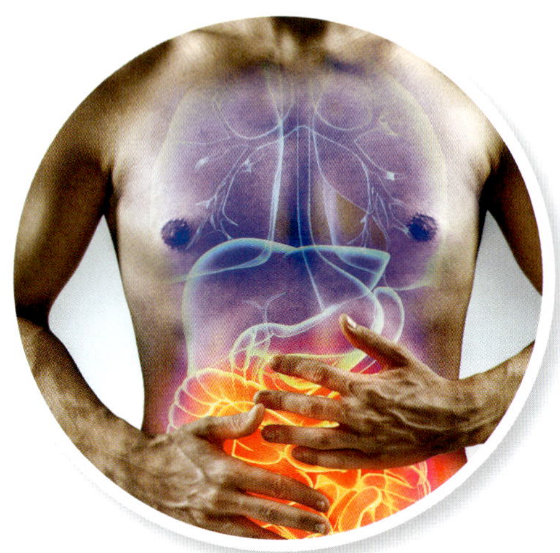

▲ *Diarrhoea causes pain, as the intestines try to shove out food that has gone bad or has bad bacteria in it*

▲ *Grains and greens are rich in fibre, which prevents constipation*

Constipation

This happens if you have not had enough water. By the time food arrives in the large intestine, it becomes very dry, making it hard for the organ to move it along. It also happens if you do not have enough fibre in your diet. Fibre stimulates the intestines to keep shoving food forward. Add 5 to your age, that is the amount of fibre (in grams) that you need every day.

Flatulence

If you have had lots of beans, you hope you do not get this in the classroom. Flatulence happens when the digestive system fills up with gas and has to throw it out, often making a whistling sound. Beans and fibre-rich foods have carbs that cannot be digested by the small intestine. But the bacteria in the large intestine can digest them, producing carbon dioxide. If your bacteria are not healthy, they might produce bad smelling gases.

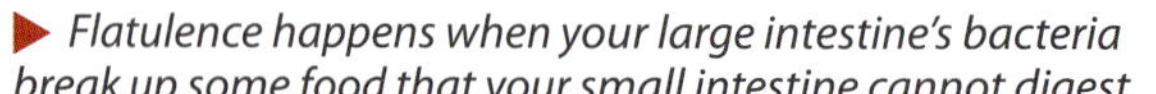

▶ *Flatulence happens when your large intestine's bacteria break up some food that your small intestine cannot digest*

The Urinary System

After your tissues have converted food into metabolites and energy, the remainder is waste. It is drained out by your blood and taken to the urinary system, which filters and removes it.

In a day, most people will pass 2,500 ml of urine and drink about that much of water. This is controlled by the brain and the pituitary gland. The brain makes sure that you do not need to pass urine when you are sleeping. The pituitary gland makes a hormone called the anti-diuretic hormone, which controls how much water you urinate, so that you do not become dehydrated.

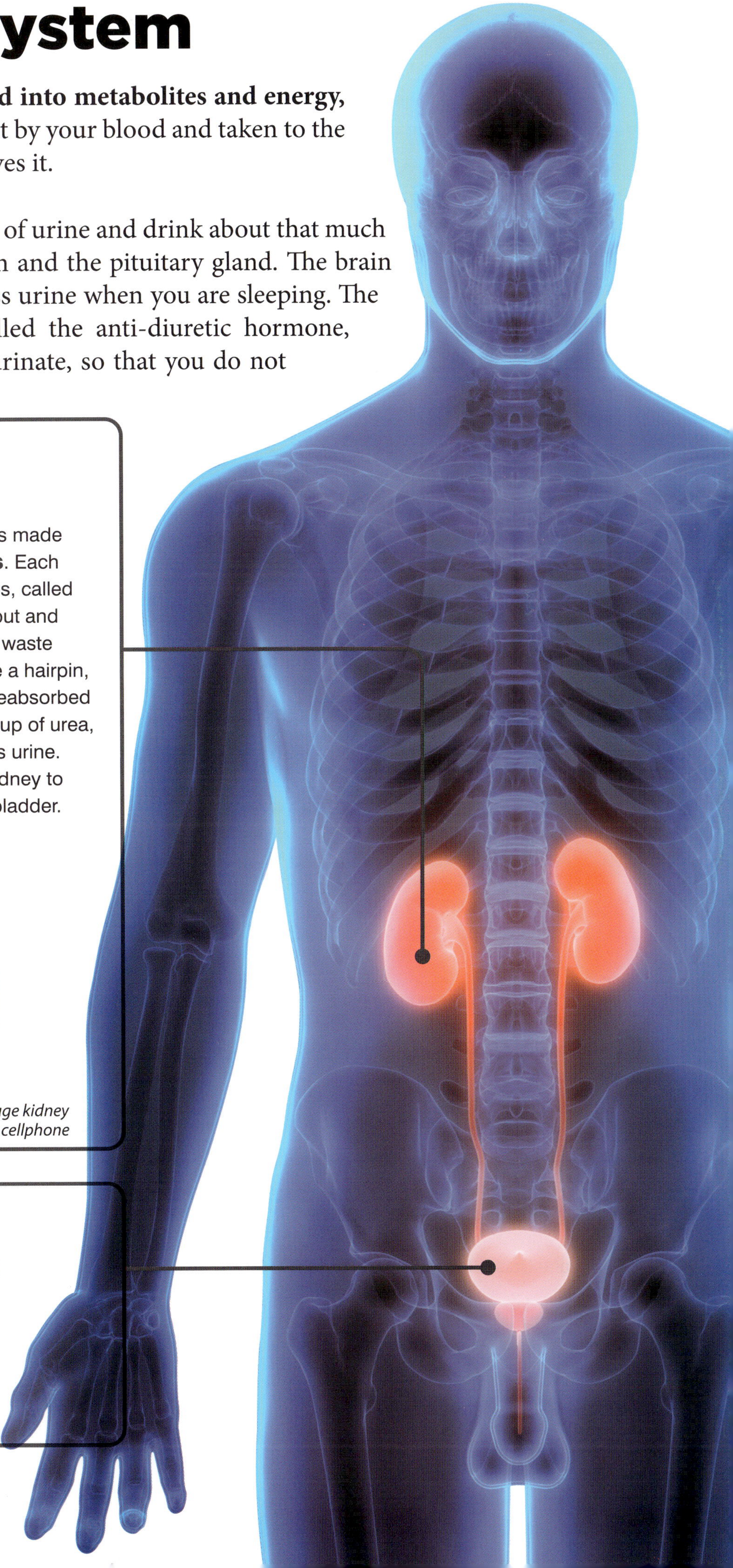

Kidneys

These are two bean-shaped organs near your intestines, which filter the blood. Each kidney is made up of thousands of tiny parts called **nephrons**. Each nephron is made of a tangle of blood capillaries, called glomerulus, which allows the plasma to seep out and enter the cup-shaped Bowman's capsule. The waste travels further through a long tube, shaped like a hairpin, the loop of Henle, where most of the water is reabsorbed by the blood. What remains finally is a thick soup of urea, other wastes, and a little water, which becomes urine. The urinary ducts merge at the centre of the kidney to become the **ureters**, which take urine to the bladder.

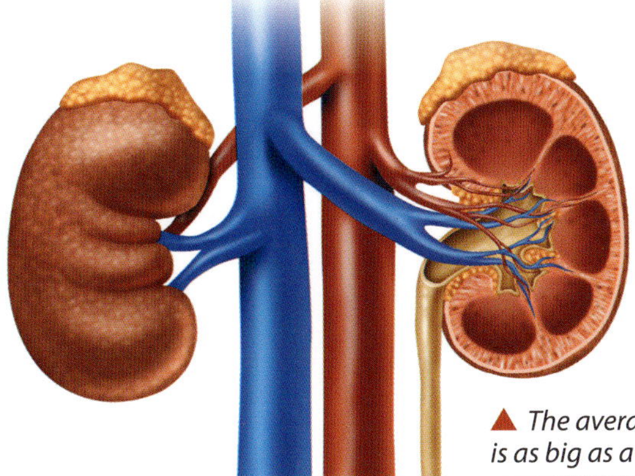

▲ *The average kidney is as big as a cellphone*

Urinary Bladder

Did you know that the urinary bladder can hold up to half a litre of water? It collects the urine from the kidneys till enough has built up. Then the nerves of the bladder tell the brain that it is time to pass urine. The urine finally passes through the urethra out into the environment.

Gall & Kidney Stones

Diseases like liver cancer, infections, or anaemia cause the gall bladder to form gallstones. Similar things happen in the kidney too, where **kidney stones** are formed. Doctors know you have got them if:

* you feel a lot of pain in your belly or your back;
* there is blood when you urinate or excrete;
* you get fever and shivering;
* your urine looks cloudy or your excreta looks pale;
* your urethra burns when you pass urine.

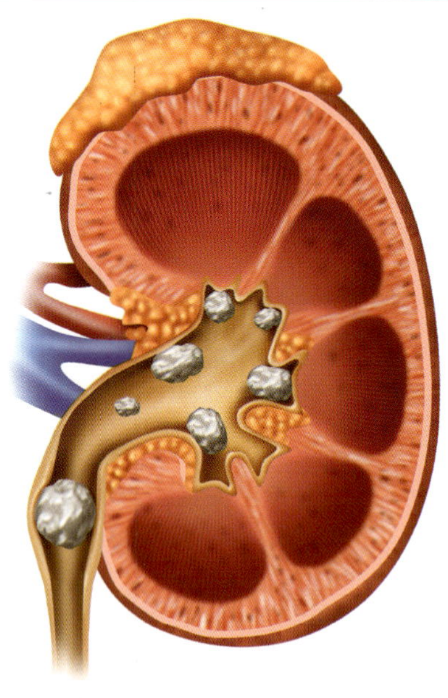

Kidney stones are more common in summer and in hotter climates

Gallstones

A gallstone is formed when bile becomes very thick due to lack of water, and the bile salts begin to crystallise. These tiny crystals then block the bile duct, causing a lot of pain and indigestion, because bile does not reach the intestine. Gallstones may often have to be removed by surgery.

Gallstones cause lasting, severe pain and indigestion

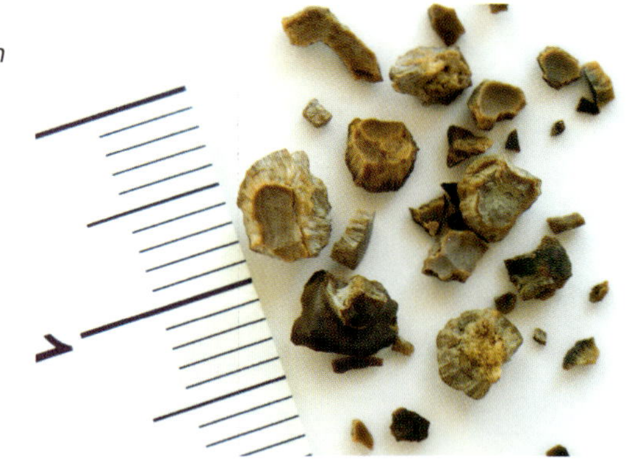

Kidney stones taken out after surgery. They can grow upto 5 mm wide

Kidney Stones

Drinking a lot of water every day usually stops kidney stones from forming. If there's too little water in the body, the salts and urea filtered from the blood into the kidney begin to crystallise. These stones block the ureters and can cause a lot of pain.

Ultrasound

Certain sounds with an ultrasonic frequency are inaudible to the human ear. Such sounds are known as ultrasound. The vibrations caused by this sound can help break up kidney stones and gall stones in the body into tiny pieces that come out easily and unblock the kidneys or gall bladder. This helps you avoid surgery.

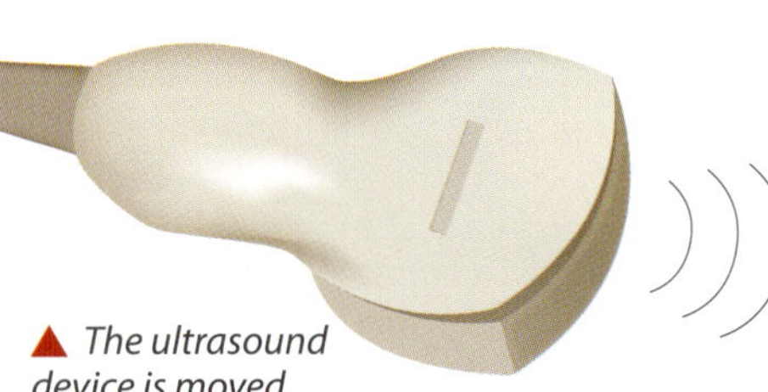

The ultrasound device is moved over the skin of the stomach

Incredible Individuals

Alexander, the Greek conqueror, died in great pain when he was just 33. Some medical historians believe that the symptoms of his disease described by the writers of his time, point to an inflammation of the gall bladder, **cholecystitis**, which is usually caused by gallstones.

Alexander of Macedon is commonly known as Alexander the Great

A Healthy Mind & Healthy Body

A diet that has the right amounts of proteins, carbs, fats, vitamins, minerals, and water is a balanced diet. However, because of stress, bad eating habits, or even mental conditions, you may fall short of one or more of these nutrients. This is called **malnutrition.**

Breakfast is Essential... Even If You're in a Hurry

Breakfast is the most important meal. When you wake up, your body has digested its dinner and both your brain and muscles need fresh energy. A good breakfast does that—it gives you the energy to stay awake and alert.

One must always try to have a hearty and balanced breakfast

Snacks are Good for You

Children eat less in meals and like having a snack now and then. It keeps your energy levels up. Fries, crisps, and chocolate are high in sugar—have a little if you've not had a meal in a long time. But the best are fruits, muffins, bread-and-butter, or nuts.

Eat Less, but Eat More

Puzzled? We mean to say, eat less in each meal, but have more meals. Research says that fewer meals make you overeat, while more meals help you eat smaller, healthier portions. A 3-ounce bag of chips actually has 3 servings, so don't eat it all at once. Share it with friends or eat it over 3 days.

Smaller portions make for healthier meals

Sugar and Sleepiness

After eating foods rich in proteins and carbohydrates, like meat and rice, we feel sleepy. This is because the sugar and amino acids entering the blood make the pancreas produce more insulin. Insulin pushes sugars and amino acids into the tissues, except for one amino acid called tryptophan. This instead goes into the brain, where it is turned into serotonin. Serotonin tells the brain to slow down and make the body go to sleep. Sleep is good for digestion as it needs energy. If you don't get good sleep, you can have problems digesting food.

A quick afternoon nap helps digest your lunch

Coffee is Good... in Small Amounts

Some adults love drinking coffee, sometimes up to 10 cups a day. The caffeine in it keeps them awake and refreshed. It helps them pass urine and maintain water balance. But it is good only as long as they drink under 400 mg of caffeine—which is about four cups of coffee—in a day. Too much caffeine causes more acid to be released in the stomach and stops your body from absorbing calcium. So if mum or dad have a lot of coffee, tell them to go easy!

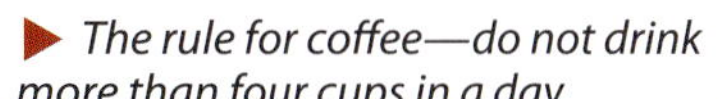

The rule for coffee—do not drink more than four cups in a day

Word Check

Aerodynamic: It means to have qualities which make it possible to move through air easily.

Alveolar macrophage: It is an immune cell in the alveoli that destroys germs.

Alveolar pressure: It is the pressure of the mix of gases in the alveoli.

Alveoli: They are the functional units of the lungs where gas exchange happens.

Aquaculture: The rearing and cultivation of aquatic animals and plants for food.

Arthropods: They form the largest order of the animal kingdom, mainly consisting of animals with 'jointed' legs.

Articular cartilage: It is the cartilage that cushions the bones in a joint.

Atoms: They are the smallest units of a chemical element.

Atria: They are the upper chambers of the heart.

Axon: Part of a neurone that takes the action potential to the next neurone

Boyle's Law: It is the law of physics that governs the relationship of a gas' volume and pressure.

Bronchial tree: It is the network of bronchi and bronchioles.

Bronchi: They are the branches of the trachea leading into the lungs.

Bronchitis: It is the inflammation of the bronchi and bronchioles leading to difficulty in breathing.

Bursitis: It is the infection of the synovial sac.

Carapace: It is the hard, upper shell structure of a mammal like a glyptodont or tortoise.

Chitinous: It refers to animals with a semi-transparent substance which forms part of the exoskeleton in arthropods.

Chordate: It refers to an animal that possesses vertebrae and belongs to the large phylum Chordata.

Collagen: It is a protein that forms threads easily. It is seen in tendons and ligaments, as well as hair and nails.

Compaction: It is the exertion of force on something so that it becomes denser.

Conchae: They are also called turbinates. They are the folds of the skull that swirl the air in your inner nose.

Coniferous: Trees that have needle- or scale-like leaves, and reproduce by forming a cone to contain seeds. Coniferous trees are evergreen, but not all evergreens have cones.

Cross pollinators: It refers to sources of pollination of a flower or plant with pollen from another flower or plant.

Dermis: It is the inner layer of the skin, made of live cells, hair follicles, and sebaceous glands.

Diaphragm: It is a giant muscle attached to the ribs, sternum, and spine that helps in breathing.

Electroreceptive organs: They are special organs in the skin of certain fish, which help them detect electric signals.

Endothermy: The ability to generate heat and regulate the body temperature internally

Epidermis: It is the outer layer of skin that acts like a waterproof wall against pathogens and allergens.

Epiglottis: A movable piece of cartilage that closes the wind pipe while swallowing food.

Eupnea: It means to breathe without making a conscious effort.

Exoskeleton: It is the hard, external covering seen in invertebrate animals.

Extrusive: It refers to rocks that have formed on Earth's surface from the lava or volcanic matter that burst out from underneath its crust.

Fault: It refers to a crack formed in the crust by the movement of tectonic plates, creating immense pressure on the crust.

Goblet cells: They are the cells in the respiratory epithelium that make mucus.

Gondwana: It is the continent that was formed in the southern hemisphere after the break-up of Pangaea. It later separated to form South America, Africa, Antarctica, Australia, Arabia, and the peninsula of India.

Grey matter: It is the part of the brain and spinal cord that is made of the cell bodies of neurones.

Gustatory receptor cells: The cells in the tongue which pick up taste and relay it to the sensory nerves.

Gyri: They are the folds of the cerebral cortex full of grey matter.

Haemoglobin: It is a special protein in the RBCs that contains iron.

Helminths: They are a class

of microscopic worms, some of which infect us.

Hormones: They are the chemical messengers that travel through blood.

Hyperpnea: It means to breathe with a conscious effort.

Hypodermis: It is the third layer of the skin. It is a deeper subcutaneous tissue made up of fat and connective tissue.

Ice crystals: Precipitation that consists of slow falling crystals of ice.

Immunoglobulin: It is the medical term for antibodies.

Inferno: It is a large fire that is dangerously out of control.

Intercostal muscles: They are the skeletal muscles between your ribs, used in breathing.

Interneurones: They are neurones in the spinal cord that control reflexes.

Intrusive: It refers to the rocks that have formed beneath Earth's crust.

Keratin: It is a fibrous protein that is present on the outer layer of feathers, nails, and hooves.

Kidney stones: They are crystals of urea or salts that form in your kidneys.

Lamina propria: It is the tissue covering the bronchi, which has BALT.

Land breeze: It is the wind moving from a landmass to the water.

Larva: It is a stage in the life of an amphibian or insect where it is completely different from its adult form.

Laryngopharynx: It is the part of the pharynx above the larynx that acts as a resonating chamber.

Latitude: It is the angular distance of a place, lying north or south of Earth's Equator, usually expressed in degrees (°) and minutes (‘), for example, the Tropic of Cancer lies 23° 26’ 21’’ N.

Ligaments: They are the collagen sheets that help with the movement of the joints.

Lingual tonsils: It is the lymphoid tissue under the tongue.

Intervertebral disc: It is a piece of cartilage that cushions vertebra from each other.

Lobe-finned fish: These fish have a central appendage made up of bones and cartilage in their fins. This is what helped tetrapods walk on land.

Lymphatic system: It helps carry fluids around the body. Therefore, it is helpful to the circulatory system. It also helps to protect our body from diseases and infections.

Macrophages: These are immune cells that eat and destroy pathogens in the body.

Magma: It is the hot fluid or semi-fluid material below or within Earth's crust.

Magnitude: It is a number that explains the intensity of an earthquake.

Malleable: It refers to materials that can be hammered into shape without breaking them.

Malnutrition: It is a disorder caused by not having enough food or certain nutrients in food.

Mandible: It is the bone of the lower jaw.

Maxilla: It is the bone of the upper jaw.

Medulla oblongata: It is the part of the brain that connects it to the spinal cord.

Meninges: It forms the protective covering of the brain and the spinal cord.

Mesozoic Era: It was a period around 65 million years ago when giant dinosaurs, birds and reptiles dominated Earth.

Meteorology: It is the branch of science that studies the atmosphere; it is very useful in predicting the weather.

Microscopic: It is used to describe something that is so tiny that it is only visible through a microscope.

Molecules: They are a group of atoms bonded together, representing the smallest fundamental unit of a chemical compound that can take part in a chemical reaction.

Monophagous: It refers to feeding on a single kind of plant or animal.

Moulting: It is the process by which an animal sheds its exoskeleton and grows another.

Moulting: It is the process by which an animal sheds its exoskeleton to regrow a new one.

Myelin sheath: It is the protective covering around axons of neurones.

Myriapods: It is an arthropod group; the animals have elongated bodies with numerous legs.

Nasal-associated lymphoid tissue (NALT): It is the lymphoid tissue in the nose that helps fight germs.

Nasopharynx: It is the part of the pharynx between the nose and the mouth.

Nephrons: They are the tiny parts of the kidney that filter blood.

Neurotransmitters: They are the chemicals in synapses that turn up or turn down neurones.

Notochord: It refers to the cartilaginous skeletal rod that supports the body of an animal when it is at the embryonic (developing) stage.

Olfactory bulb: It is the part of the brain that deals with smell.

Olfactory receptor cells: They are the cells in your nose that catch smelly chemicals.

Omnivorous: It refers to an animal that eats both plants and other animals.

Orbit: It is the socket in the skull where the eye sits.

Oropharynx: It is the part of the pharynx where the food pipe and the wind pipe cross.

Osteoblasts: They are cells that make collagen.

Osteoclasts: These are the cells that resorb the bone's matrix to remodel it.

Oxyhaemoglobin: It is the haemoglobin that is chemically attached to oxygen.

Palatine tonsils: It is the lymphoid tissue in the oropharynx.

Papillae: They are the taste organs in your tongue.

Paranasal sinuses: They are the hollow, air-filled spaces in the skull that help clean air.

Pathogens: They are the germs that attack us, like bacteria, fungi, viruses, protozoa, and helminths.

Pharyngeal tonsils: Also called adenoids, it is the lymphoid tissue in the nasopharynx.

Photoreceptor cells: They are the cells in the retina that are sensitive to light, made up of rod cells and cone cells.

Photosynthesis: It is the process by which green plants prepare food in the presence of sunlight and a pigment called chlorophyll that is present in the leaves of plants. The plants also need carbon dioxide and water to release energy and oxygen.

Pituitary gland: It is the gland that makes many hormones, it is controlled by the hypothalamus.

Pleura: It is the bag that cushions the lungs.

Pulmonary arteries: They are the arteries that take deoxygenated blood from the heart to the lungs.

Pulmonary lobes: They are the internal divisions of the left and right lungs, further divided into segments, lobules, and alveolar sacs.

Pulmonary surfactant: It is a soap-like liquid in the lungs that helps diffuse oxygen.

Pulmonary veins: They are the veins that take oxygenated blood from the lungs to the heart.

Pyroclastic flows: It is a current of volcanic matter and hot gases that quickly gushes out from a volcano at speeds of 100–700 kmph.

Respiratory bronchiole: It is the part of the bronchiole that ends in an alveolus.

Respiratory epithelium: It is the lining of the conducting zone of your respiratory system.

Respiratory zone: It is the part of your respiratory system that exchanges gases with blood.

Rod cells: They are the cells in the retina that sense whether it is light or dark.

Sauropsids: It is a group of amniotic animals that consists of birds and reptiles.

Schwann cells: They are the cells that make myelin and cover the axons of neurones.

Sea breeze: It is the wind moving from water to a landmass.

Serotonin: It is the chemical in your brain that induces sleep.

Subduction: It is a situation when one tectonic plate in Earth's crust is forced under another.

Subduction: It refers to the downward and sideways movement of Earth's crust into the mantle, beneath another plate.

Sublimation: It is a process where a solid turns into gas without being converted into the liquid form.

Tendons: It is a tissue that attaches muscle to the bone.

Tetrapod: It is an animal that possesses four feet.

Topography: It refers to the study of how natural structures are arranged in a certain part of land.

Trachealis muscle: It is the muscle that contracts the trachea.

Transpiration: It is the exhalation of water vapour through leaves.

Triassic takeover: It lasted between 251 and 199 million years ago after the Permian Mass Extinction wiped out most synapsids and sauropsids began to evolve.

Ureters: Tubes that take urine from the kidneys to the urinary bladder.

Uvula: It is a flap of tissue that stops food going up into the inner nose.

Ventricles: They are the lower chambers of the heart.

Vertebra: It is a unit of the vertebral column.

Vertebrates: Animals including mammals, fish, amphibians, birds, and reptiles that possess a backbone or spinal column.

Viviparous: It refers to a group of animals that reproduce by live birth instead of reproducing by laying eggs.

White matter: It is the part of the brain and spinal cord made of the axons and dendrites of neurones.

Image Credits

Cover

From left to right

Shutterstock: Front: Axel Alvarez; BlueRingMedia; Nathapol Kongseang; tr/Yellow Cat; vladsilver; panor156; wacomka; GraphicsRF.com; Sarunyu_foto; Alex Staroseltsev; Andrew Burgess; Eric Isselee; DM7; Sebastian Janicki; New Africa; Jiang Zhongyan; Grigorev Mikhail; Kletr; Vac1; robert_s; Inachis Projekt; Back: Romolo Tavani; Bonita R. Cheshier; Tatyana Mi; Radek Borovka; Dennis W Donohue; Vibe Images; Sunti; Anan Kaewkhammul; FrentaN; Hintau Aliaksei; Smit; Anton-Burakov; Eric Isselee; lusia83; Rattachon Angmanee; Eric Isselee; Quang Ho; Eric Isselee; design36

a: above, b: below/ bottom, c: centre, f: far, l: left, r: right, t: top, bg: background

Our Planet Earth

Shutterstock: 7b/everst; 8tr/Drp8; 8b/melissamn; 9tr/MBL1; 9tl/Romolo Tavani; 9c/Brian B Fox; 9bl/Galyna Andrushko; 9br/Jose F. Donneys; 10t/Sean Pavone; 10b/meganopierson; 11tr/Creative Jen Designs; 11cl/Daniel Prudek; 11c/Lucky-photographer; 11cr/Dmitry Rukhlenko; 11br/IKO-studio; 11b/hrui; 12tr/Designua; 12fr/Martin Mecnarowski; 12b/finchfocus; 12&13bg/sittitap; 13cl/Evgeny Haritonov; 13cr/RunArt; 13br/Erdal Akan; 14bg/Eakkarut Choeichin; 14tr/aappp; 14cr/dugdax; 15b/Konstantin Grosch 1986; 16tbg/06photo;16tbg/GoodFocused; 16bbg/JaySi; 16b/LutsenkoLarissa; 17bbg/David Steele; 18tr/Ekkachai;18c/ESB Professional ; 18br/Rich Carey; 18&19bg/Ema Martin; 19t/Odua Images; 19t/Rawpixel.com; 20br/Allen.G; 20bl/Andrei Armiagov; 20&21c/titoOnz; 21tr/Incredible Arctic; 21br/zlikovec; 21b/CherylRamalho; 22tr/H.Tanaka; 22c/divedog; 22bl/blue-sea.cz; 22&23bg/Vlad61; 23tr/weera sreesam; 23c/Tory Kallman; 23br/miha de; 23b/Nicole Helgason; 24c/peipeiro; 24bl/Elfangor; 24br/Rattiya Thongdumhyu; 25tr/Vladislav Ogorodnyk; 25cr/Lynn Yeh; 25br/Jasper Suijten; 26cl/Gerrit Bunt; 26cr/Claudio Caridi; 27cl/suronin; 27cr/Ian Winslow; 28t/sirtravelalot; 28b/Matauw; 28&29bg/KSPTT; 29tr/Richard Bradford; 29cl/Drp8; 30cl/Mr. SUTTIPON YAKHAM; 30&31bg/Willyam Bradberry; 31tr/Zigzag Mountain Art; 31b/Smit; 32tr/Fokin Oleg; 32br/Fokin Oleg; 32cb/losmandarinas; 32br/Breck P. Kent; 32&33bg/Leene; 33tr/Skynavin; 34tr/Sergejus Lamanosovas;34c/vilax; 34b/Rattachon Angmanee; 35tr/LesPalenik; 35cr/Mauro Rodrigues; 35b/Teresa Otto; 36tr/FabrikaSimf; 36br/snapgalleria; 37t/lambretta_27; 37tb/Sarah2; 37c/EMJAY SMITH; 37b/Sallehudin Ahmad; 37bb/MemoPlus; 38cr/Aleks Kvintet; 38&39bg/Signature Message; 39cl/Everett Collection; 39br/Earl D. Walker; 40tr/Branko Jovanovic; 40cl/Dan Olsen; 40br/Vasiliy Ptitsyn; 41tr/Ihor Matsiievskyi; 41c/Fokin Oleg; 41br/sasimoto; 41b/Meowcyber; 42tr/Roy Palmer; 42cl/sumire8; 42fr/Bjoern Wylezich; 42bl/DestinaDesign; 42br/A_Dozmorov; 43tl/Aleksandr Pobedimskiy; 43tr/Alessandro Zappalorto; 43cl/Chatrawee Wiratgasem; 43cr/Aleksandr Pobedimskiy; 43cl/Delennyk; 43cr/Besjunior; 43bl/Dmitry Chulov; 43br/Albert Russ; 44t/photoNN; 44b/gnoparus; 44bg/Romolo Tavani; 45tl/Catmando; 45fr/Benny Marty; 45b/SL-Photography; 46cl/AvDe; 46br/Kokophotos; 46&47c/Aldona Griskeviciene; 47t/Misbachul Munir; 47cr/Vitalii Gaidukov; 48t/Alexyz3d; 48tr/TTstudio; 48fr/Fotos593; 49cl/Dariush M; 49cr/Mistervlad; 49bl/fboudrias; 49br/T photography; 50tl/Thomas Bethge; 50tr/Mopic; 50fr/d1sk; 50bl/Barks; 50b/d1sk; 51tr/Naeblys; 51/Cico; 51VectorMine; 52cr/Everett Collection; 52b/Sergey Nivens; 53tl/Hung Chung Chih; 53tr/D.Cz.; 53bl/polkadot_photo; 53br/Stickerologia; 54;55c/traXX; 54cl/Tom Wang; 54br/Bonita R. Cheshier; 54bc/Blamb; 55cr/Vitaliy Kaplin; 55b/Jim Lopes; 56b/Bigmouse108; 57cr/udaix; 57br/ Rudra Narayan Mitra; 57b/Sergey Novikov; 58tl/FoxGrafy; 58tr/FoxGrafy; 58c/Drp8; 58bl/Ondrej Prosicky; 58br/Sergey 402; 59t/CkyBe; 59cl/Rene Holtslag; 59cr/SKYEX DRONE; 60tr/Art Painter; 60cr/apiguide; 60bl/Brett Andersen; 60b/savenkov; 61tr/Designua; 61cr/LedyX; 61cl/gorillaimages; 61br/deakins83; 61b/Standret; 62tr/Designua; 62c/udaix; 63cl/VovanIvanovich; 64tr/Vaclav Mach; 64cr/solosergio; 64bl/Paul Fleet;

Wikimedia Commons: 14br/Amur Tiger Panthera tigris altaica Cub Walking 1500px.jpg/Photo by and (C)2007 Derek Ramsey (Ram-Man). Curves adjustment by Lycaon on 08:30; July 20; 2007.; GFDL 1.2 <http://www.gnu.org/licenses/old-licenses/fdl-1.2.html>; via Wikimedia Commons/wikimedia commons; 15tr/22tr/File:Distribution Taiga.png//wikimedia commons; 16bl/File:Controlled burn in grassland (5474620598).jpg/U. S. Fish and Wildlife Service - Northeast Region; Public domain; via Wikimedia Commons/wikimedia commons; 17t/File:Deserts.png/Emilfaro at English Wikipedia; Public domain; via Wikimedia Commons/wikimedia commons; 17br/File:Guatiza - Jardín de Cactus - Lanzarote - J30.jpg/Luis Miguel Bugallo Sánchez (Lmbuga); CC BY-SA 3.0 <https://creativecommons.org/licenses/by-sa/3.0>; via Wikimedia Commons/wikimedia commons; 17bl/File:Camelface.jpg/Henryhbk; Public domain; via Wikimedia Commons/wikimedia commons; 19b/File:Comet Hartley 2.jpg; via Wikimedia Commons/wikimedia commons; 17bl/File:Camelface.jpg/Henryhbk; Public domain; via Wikimedia Commons/wikimedia commons; 26tl/File:N-Mesopotamia and Syria english.svg/via Wikimedia Commons/wikimedia commons; 27tr/File:Indus Valley Civilization; Mature Phase (2600-1900 BCE).png; via Wikimedia Commons/wikimedia commons; 29b/File:Sir Henry Raeburn - James Hutton; 1726 - 1797. Geologist - Google Art Project.jpg; via Wikimedia Commons/wikimedia commons; 30tr/File:Morceaux de granite superposé a à Parakou-Bénin.jpg; via Wikimedia Commons/wikimedia commons; 40bl/File:Friedrich Mohs.jpg; via Wikimedia Commons/wikimedia commons; 41bl/File:Jöns Jacob Berzelius.jpg; via Wikimedia Commons/wikimedia commons; 48c/File:Fourpeaked-fumaroles-cyrus-read1.JPG; via Wikimedia Commons/wikimedia commons; 48br/File:KatiaMauriceKrafft.jpg; via Wikimedia Commons/wikimedia commons; 49tr/File:Mount Tambora Volcano; Sumbawa Island; Indonesia.jpg; via Wikimedia Commons/wikimedia commons; 51b/File:CharlesRichter.jpg/ via Wikimedia Commons/wikimedia commons;

Animals

Shutterstock: 65cr/Nico.Stock; 65b/Piyaset; 67cl/Kurit afshen; 67b/yod 67; 68tr/Warpaint; 68cl/Herschel Hoffmeyer; 68;69b/Michael Rosskothen; 69t/AKKHARAT JARUSILAWONG; 69cr/Michael Rosskothen; 70bg/luck luckyfarm; 70-71/Eric Isselee; 70bl/WildlifeWorld; 71tr/ehtesham; 71bl/Narupon Nimpaiboon; 72l/FloridaStock; 72bl/Daniel_Kay; 73tr/Mike Truchon; 73tl/Eric Isselee; 73cr1/Tracy Starr; 73cr2/Four Oaks; 73bl/Eric Isselee; 73br/Wirestock Creators; 74tl/TobyG; 74cr/nadtytok; 74bl/irin-k; 74bc/Jesse Nguyen; 74br/Rosa Jay; 75tr/tristan tan; 75cl/Kerry Hargrove; 75br/photomaster; 76t/Patthana Nirangkul; 76cl/Javier Valladares; 76cr/Maciej Olszewski; 76bl/Francois Loubser; 77bl/colacat; 77bc/Christian Musat; Anastasiia Petrych; 77br/Krakenimages.com; 78bl/tienduc1103; 78-79c/anat chant; 79tr/Radu Bercan; 79bc/Luc Pouliot; 80tr/Brett Hondow; 80cr/Mr. SUTTIPON YAKHAM; 80bl/Tomasz Klejdysz; 80br/Achira22; 80tr/BOONCHUAY PROMJIAM; 80b/Leena Robinson; 80br/Protasov AN; 82tr/JD Phote; 82cl/PetrP; 82b/Eric Isselee; 83tr/Ondrej Prosicky; 83b/Evgeniy Ayupov; 84tr/paula french; 84b/Tsekhmister; Ortis; 85trAndrey Pavlov; 85bl/Achkin; 86tr/Benny Marty; 86cl/Cheng Wei; 86br/Eugen Thome; 87tr/Esteban De Armas; 87cl/Aunt Spray; 88cl/Roblan; 88b/Michal Hykel; 89tr/Kristine Reed; 90tr/cbpix; 90cl/Dr Morley Read; 90bl/PetlinDmitry; 90br/Daleen Loest; 91tr/Teguh Tirtaputra; 92bl/Henrik Larsson; 92br/NatureDiver; 92b/Henk Bogaard; 93tr/ppl; 93br/piya; saisawatdikul; 93bl/Witaya Proadtayakogool; 94tr/Process; 94br/Luca Santilli; 95cr/Jan Danek jdm.foto; 95br/Protasov AN; 96bg/Nneirda; 96tr/Neil Bromhall; 96bl/ILYA AKINSHIN; 97tr/Yann hubert; 97cl/Kirsanov Valeriy Vladimirovich; 97br/bmszealand; 98tr/65cr/Nico.Stock; 65b/Piyaset; 67cl/Kurit afshen; 67b/yod 67; 68tr/Warpaint; 68cl/Herschel Hoffmeyer; 68;69b/Michael Rosskothen; 69t/AKKHARAT JARUSILAWONG; 69cr/Michael Rosskothen; 70bg/luck luckyfarm; 70-71/Eric Isselee; 70bl/WildlifeWorld; 71tr/ehtesham; 71bl/Narupon Nimpaiboon; 72l/FloridaStock; 72bl/Daniel_Kay; 73tr/Mike Truchon; 73tl/Eric Isselee; 73cr1/Tracy Starr; 73cr2/Four Oaks; 73bl/Eric Isselee; 73br/Wirestock Creators; 74tl/TobyG; 74cr/nadtytok; 74bl/irin-k; 74bc/Jesse Nguyen; 74br/Rosa Jay; 75tr/tristan tan; 75cl/Kerry Hargrove; 75br/photomaster; 76t/Patthana Nirangkul; 76cl/Javier Valladares; 76cr/Maciej Olszewski; 76bl/Francois Loubser; 77bl/colacat; 77bc/Christian Musat; Anastasiia Petrych; 77br/Krakenimages.com; 78bl/tienduc1103; 78-79c/anat chant; 79tr/Radu Bercan; 79bc/Luc Pouliot; 80tr/Brett Hondow; 80cr/Mr. SUTTIPON YAKHAM; 80bl/Tomasz Klejdysz; 80br/Achira22; 80tr/BOONCHUAY PROMJIAM; 80b/Leena Robinson; 80br/Protasov AN; 82tr/JD Phote; 82cl/PetrP; 82b/Eric Isselee; 83tr/Ondrej Prosicky; 83b/Evgeniy Ayupov; 84tr/paula french; 84b/Tsekhmister; Ortis; 85trAndrey Pavlov; 85bl/Achkin; 86tr/Benny Marty; 86cl/Cheng Wei; 86br/Eugen Thome; 87tr/Esteban De Armas; 87cl/Aunt Spray; 88cl/Roblan; 88b/Michal Hykel; 89tr/Kristine Reed; 90tr/cbpix; 90cl/Dr Morley Read; 90bl/PetlinDmitry; 90br/Daleen Loest; 91tr/Teguh Tirtaputra; 92bl/Henrik Larsson; 92br/NatureDiver; 92b/Henk Bogaard; 93tr/ppl; 93br/piya; saisawatdikul; 93bl/Witaya Proadtayakogool; 94tr/Process; 94br/Luca Santilli; 95cr/Jan Danek jdm.foto; 95br/Protasov AN; 96bg/Nneirda; 96tr/Neil Bromhall; 96bl/ILYA AKINSHIN; 97tr/Yann hubert; 97cl/Kirsanov Valeriy Vladimirovich; 98tr/bmszealand; 98bl/Warpaint; 98bg/Aliaksandr Antanovich; 99tr/Daniel Eskridge; 99cl/Catmando/ 100&101bg/; 100&101 center/Nicolas Primola; 102tr/Tory Kallman; 102b/A_Mikhail; 103tr1/Anan Kaewkhammul; 103tr2/Gino Santa Maria; 103cl/Ondrej Prosicky; 103bl/Ortis; 103br/Eric Isselee; 104cl/Gino Santa Maria; 104cr/apple2499; 104bl/nrey; 104br/Ortis, Olis Design; 105tr/; 105cl/Sergei25; 105br/Eric Isselee; 106tr/Catmando; 106cl/Oleg Lopatkin; 106br/solar22; 107cl/Warpaint; 107cr/Warpaint; 108tr/funny face; 108&109c/photomaster ; 108br/pfluegler-photo; 109tr/Gunnar Pippel; 109bl/ Napat; 110c/Vangert; 110bl/mnoor; 111c/Konstantin G; 111bl/Daniel Huebner; 111br/Pierro Pozella; 112tr/Tomas Kotouc; 112&113c/frantisekhojdysz, 113tr/Kletr; 113br/wildestanimal; 114tr/Matt9122; 114cr/FtLaud; 114bl/Anita Kainrath; 115tr/Yellow Cat; 115c/pan demin; 116b/GraphicsRF.com; 116cl/GIBAN; 116c/Warpaint; 118b/Cathy Keifer; 119tr/Tatiana Diuvbanova; 119cr/JIANG HONGYAN; 119c/Suwat wongkham; 119bl/Margaret M Stewart; 120bl/Catmando; 121cl/Ekaterina Glazkova; 112cr/Michael Rosskothen; 121cl/Herschel Hoffmeyer; 121bl/Michael Rosskothen; 122&123c/nattanan726; 123tr/seasoning_17; 123c/Kuttelvaserova Stuchelova; 124tr/reptiles4all; 124cl/DWI YULIANTO; 124cr/Valt Ahyppo; 124bl/fivespots; 125tl/Dave Montreuil; 125cr/Rebecca Harrison; 125cl/Michiel de Wit; 125tr/IrinaK;

Wikimedia Commons: 86bl/File:Opabinia smithsonian.JPG/wikimedia commons; 87cr/File:Pterygotus anglicus reconstruction.jpg/wikimedia commons; 87br/File:Arthropleura NT small.jpg/wikimedia commons; 106bl/File:Acanthodes BW spaced.jpg/wikimedia commons; 107b/File:Helicoprion bessonovi1DB.jpg/wikimedia commons; 116tr/File:Slimonia acuminata 1.jpg/wikimedia commons; 117tr/File:Skeleton of Ichthyostega.JPG/wikimedia commons; 120tr/File:Hylonomus BW.jpg/wikimedia commons; 120br/File:Proterosuchus BW.jpg/wikimedia commons;

Human Body

Shutterstock: 127b/patrice6000; 128l/SciePro; 128cr/bogadeva1983; 128br/hidesy & itor; 129tr/Christos Georghiou; 129cl/Sakurra; 129bl/Simona Chira; 129br/Sirintra Pumsopa; 130cl/decade3d - anatomy online; 130bl/stihii; 130&131c/ilusmedical; 131cl/marina_ua; 131cr/AlexLMX; 132tl/Daniel Eskridge; 132cr/gritsalak karalak; 132cl/Designua; 132bl/Sari Oneal; 132br/Sari Oneal; 133tl/pablofdezr; 133cr/solar22; 133bl/AlexanderBee; 134&135c/Alex Mit; 134bl/Designua; 134br/Christina Li; 135cr/extender_01; 136bl/Anna L. e Marina Durante; 137tr/ducu59us; 137br/decade3d - anatomy online; 137bc/Holger Kirk; 138tr/CLIPAREA I Custom media; 138bl/ yomogi1; 139tr/Africa Studio; 140bl/Lightspring 140&141c/Lyudmyla Ishchenko; 141cr/Marochkina Anastasiia; 142tl/Elena Paletskaya; 142tr/NelaR; 142br/Designua; 143tr/Kateryna Kon; 144t/Interior Design; 144cl/Standard Studio; 145tl/cenksns; 145b/paulista; 146bl/Anatomy Image; 146cl/Anatomy Image; 146c/S K Chavan; 147tl/Magic mine; 147cl/Nerthuz; 148bc/Alila Medical Media; 147br/Anatomy Image; 148tr/Jose Luis Calvo; 148br/Creations; 149tr/sciencepics; 149cl/sciencepics; 149crBlueRingMedia; 149bl/Timonina; 149br/O2creationz; 150tl/Liya Graphics; 150cr/Tatiana Shepeleva; 150bl/Kateryna Kon; 150&151/KK Tan; 151tr/Design_Cells; 151cr/Christoph Burgstedt;

151bl/Kateryna Kon; 151br/nobeastsofierce; 152&153/Sky vectors; 152tl/Vladimir Zotov; 152c/ibreakstock; 152&153tc/Kateryna Kon; 153c/Kateryna Kon; 153c/Mirror-Images; 153b/sciencepics; 153br/Karel Gallas; 154tr/Africa Studio; 154cl/Joa Souza; 154cr/Africa Studio; 154bl/Rost9; 155tr/Serhiy Kobyakov; 155cl/Rawpixel.com; 155bc/VaLiza; 156tr/Tursk Aleksandra; 156b/Alila Medical Media; 157cl/YAKOBCHUK VIACHESLAV; 157bl/Designua; 157br/Reimar; 158tr/komokvm; 158cr/stockshoppe; 158br/Blamb; 159tr/Alex Mit; 159cr/Aldona Griskeviciene; 159bl/Sakurra; 159br/Lightspring; 160cr/Vecton; 160bl/u3d; 161tr/Sakurra; 161br/Dirk Ercken; 162tr/Pete Pahham; 162cr/VectorMine; 162cl/Blamb; 162br/fizkes; 163cl/gritsalak karalak; 163br/Ground Picture; 164tr/Master1305; 164cr/VectorMine; 164b/Umomos; 165tr/Designua; 165c/VectorMine; 165br/freeskyline; 166tr/Arcady; 166bc/Sakurra; 166cr/Sathit; 166&167c/design36; 168tl/Corona Borealis Studio; 168cr/Drp8; 169tr/Prostock-studio; 169cr/Ellen Bronstayn; 169br/lovepetch; 168&169b/Real Sports Photos; 170l/Alex Mit; 170cr/Designua; 170br/eranicle; 171tr/sciencepics; 170&171/csciencepics/ 172cr/Fer Gregory; 172br/Anan Kaewkhammul; 173br/Designua; 172&173c/Vac1; 174tr/SOK Studio; 174c/joshya; 175cl/Blamb; 175br/fizkes; 174&175bc/Choksawatdikorn; 176tr/Sebastian Kaulitzki; 176cr/TimeLineArtist; 176br/Nicolas Primola; 177l/Sebastian Kaulitzki; 178tr/metamorworks/nobeastsofierce; 178cr/Akadet Guy; 179tr/Magic mine; 179cr1/Elen Bushe; 179cr2/Nerthuz; 179bl/Alila Medical Media; 180r/Sebastian Kaulitzki; 181tr/Magic mine; 181cl/Olena758; 181bl/Kateryna Kon; 18br/decade3d - anatomy online; 182tr/Romariolen; 182cl/marilyn barbone; 182br/Shift Drive; 183r/Magic mine; 183cl/Lightspring; 184tr/Lightspring; 184cl/BlurryMe; 184cr/Evan Lorne; 184br/giannimarchetti; 184bl/joshya; 185tr/CKP1001; 185cr/BW Folsom; 185cl/StockImageFactory.com;

Dreamstime: 139cr/Inna Volodina; 139br/Evgeniia Kuzmich; 185br/Photosvit

Wikimedia Commons: 86bl/File:Opabinia smithsonian.JPG; via Wikimedia Commons/wikimedia commons; 87cr/File:Pterygotus anglicus reconstruction.jpg; via Wikimedia Commons/wikimedia commons; 87br/File:Arthropleura NTsmall.jpg; via Wikimedia Commons/wikimedia commons; 106bl/File:Acanthodes BW spaced.jpg; /wikimedia commons; 107b/File:Helicoprion bessonovi1DB.jpg/wikimedia commons; 116tr/File:Slimonia acuminata 1.jpg/wikimedia commons; 117tr/File:Skeleton of Ichthyostega.JPG/wikimedia commons; 120tr/File:Hylonomus BW.jpg/wikimedia commons; 120br/File:Proterosuchus BW.jpg/wikimedia commons; 131bl/File:Moniz.jpg/wikimedia commons; 135br/File:Santiago Ramón y Cajal (1852-1934) portrait (restored).jpg/wikimedia commons; 157tr/File:Kleopatra-VII.-Altes-Museum-Berlin1.jpg/wikimedia commons; pg 178br/File:Motzume.jpg/wikimedia commons;